TECHNOLOGY GUIDE

TO ACCOMPANY

UNDERSTANDING BASIC STATISTICS
THIRD EDITION
BRASE/BRASE

Charles Henry Brase
Regis University

Corrinne Pellillo Brase
Arapahoe Community College

Laurel Tech Integrated Publishing Services

HOUGHTON MIFFLIN COMPANY   BOSTON NEW YORK

Sponsoring Editor: Lauren Schultz
Associate Editor: Jennifer King
Editorial Assistant: Kasey McGarrigle
Senior Manufacturing Coordinator: Marie Barnes
Senior Marketing Manager: Ben Rivera

Printed in the U.S.A.

ISBN: 0-618-33360-6

3 4 5 6 7 8 9 – CRS – 07 06 05

Contents

Part II: MINITAB Guide

Preface

The use of computing technology can greatly enhance a student's learning experience in statistics. This *Technology Guide* provides basic instruction, examples, and lab activities for three different tools:

TI-83 Plus
MINITAB
ComputerStat

The TI-83 Plus is a versatile, widely-available graphing calculator made by Texas Instruments. This guide shows how to use its statistical functions, including plotting capabilities.

MINITAB is a statistics software package suitable for solving problems. It can be packaged with the text. Contact Houghton Mifflin for details regarding price and platform options.

ComputerStat is an interactive, user-friendly computer software package designed to accompany *Understandable Statistics* and *Understanding Basic Statistics*. It can be used to work problems that would otherwise be handled with a calculator or pencil and paper, but ComputerStat handles larger data sets with relative ease. Institutions that adopt either text may have a complimentary license to use the software. ComputerStat is available on the new hybrid HM StatPass CD-ROM for both Windows and Macintosh platforms. This CD is packaged for free with every new textbook.

The lab activities that follow accompany the text *Understanding Basic Statistics,* 3rd edition by Brase and Brase.

In addition, over one hundred data files from referenced sources are described in the Appendix. These data files are available on the HM StatPass CD-ROM or via download from the Houghton Mifflin Web site

http://math.college.hmco.com/students

PART I

TI-83 Plus Graphing Calculator Guide

for

Understanding Basic Statistics, Third Edition

CHAPTER 1 GETTING STARTED

ABOUT THE TI-83 PLUS GRAPHING CALCULATOR

Calculators with built-in statistical support provide tremendous aid in performing the calculations required in statistical analysis. The Texas Instruments TI-83 Plus graphing calculator has many features that are particularly useful in an introductory statistics course. Among the features are

(a) Data entry in a spreadsheet-like format

The TI-83 Plus has six columns (called lists L_1, L_2, L_3, L_4, L_5, and L_6) in which data can be entered. The data in a list can be edited, and new lists can be created by doing arithmetic using existing lists.

Sample Data Screen

L1	L2	L3	1
11.2	8.1	1	
8.7	6.2	2	
6.8	3.6	-1	
3.2	.9	2	
4.5	1	1	
------	------	1	
		5	

$$L1(1)=11.2$$

(b) Single-variable statistics: mean, standard deviation, median, maximum, minimum, quartiles 1 and 3, sums

(c) Graphs for single-variable statistics: histograms, box-and-whisker plots

(d) Estimation: confidence intervals using the normal distribution, or Student's t distribution, for a single mean and for a difference of means; intervals for a single proportion and for the difference of two proportions

(e) Hypothesis testing: single mean (z or t); difference of means (z or t); proportions, difference of proportions, chi-square test of independence; two variances; linear regression, one-way ANOVA

(h) Two-variable statistics: linear regression

(i) Graphs for two-variable statistics: scatter diagrams, graph of the least-squares line

In this *Guide* we will show how to use many features on the TI-83 Plus graphing calculator to aid you as you study particular concepts in statistics. **Lab Activities** coordinated to the text *Understanding Basic Statistics* are also included.

USING THE TI-83 PLUS

The TI-83 Plus has several functions associated with each key. All the yellow items on the keypad are accessed by first pressing the yellow [2nd] key. The items in green are accessed by first pressing the green [ALPHA] key. The four arrow keys [▼][▲][►][◄] enable you to move through a menu screen or along a graph.

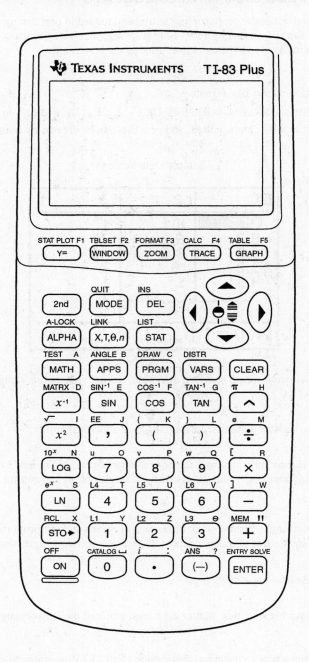

The calculator uses screen menus to access additional operations. For instance, to access the statistics menu, press ⌷STAT⌷. Your screen should look like this:

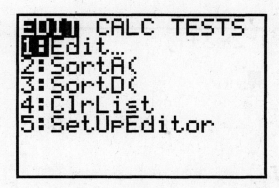

Use the arrow keys to highlight your selection. Pressing ⌷ENTER⌷ selects the highlighted item.

To leave a screen, press either ⌷2nd⌷ **[QUIT]** or ⌷CLEAR⌷, or select another menu from the keypad.

Now press ⌷MODE⌷. When you use your calculator for statistics, the most convenient settings are as shown.

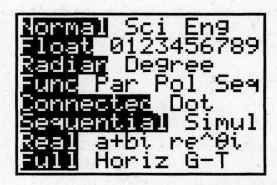

You can enter the settings by using the arrow keys to highlight selections and then pressing ⌷ENTER⌷.

COMPUTATIONS ON THE TI-83 PLUS

In statistics, you will be evaluating a variety of expressions. The following examples demonstrate some basic keystroke patterns.

Example Evaluate −2(3) + 7.

Use the following keystrokes: ⌷(-)⌷ ⌷2⌷ ⌷×⌷ ⌷3⌷ ⌷+⌷ ⌷7⌷ ⌷ENTER⌷. The result is 1.

To enter a negative number, be sure to use the key $\boxed{(-)}$ rather than the blue subtract key. Notice that the expression **-2*3 + 7** appears on the screen. When you press $\boxed{\text{ENTER}}$, the result **1** is shown on the far right side of the screen.

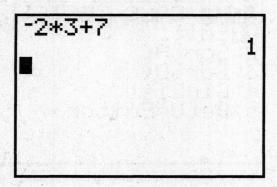

Example

(a) Evaluate $\dfrac{10-7}{3.1}$ and round the answer to three places after the decimal.

A reliable approach to evaluating fractions is to enclose the numerator in parentheses, and if the denominator contains more than a single number, enclose it in parentheses as well.

$$\frac{10-7}{3.1} = (10 - 7) \div 3.1 \quad \text{Use the keystrokes}$$

$\boxed{(}\ \boxed{1}\ \boxed{0}\ \boxed{-}\ \boxed{7}\ \boxed{)}\ \boxed{\div}\ \boxed{3}\ \boxed{.}\ \boxed{1}\ \boxed{\text{ENTER}}$.

The result is .9677419355, which rounds to .968.

(b) Evaluate $\dfrac{10-7}{\frac{3.1}{2}}$ and round the answer to three places after the decimal.

Place both numerator and denominator in parentheses: $(10 - 7) \div (3.1 \div 2)$.
Use the keystrokes

$\boxed{(}\ \boxed{1}\ \boxed{0}\ \boxed{-}\ \boxed{7}\ \boxed{)}\ \boxed{\div}\ \boxed{(}\ \boxed{3}\ \boxed{.}\ \boxed{1}\ \boxed{\div}\ \boxed{2}\ \boxed{)}\ \boxed{\text{ENTER}}$.

The result is 1.935483871 or 1.935 rounded to three places after the decimal.

Example

Several formulas in statistics require that we take the square root of a value. Note that a left parenthesis, (, is automatically placed next to the square root symbol when $\boxed{\sqrt{}}$ is pressed. Be sure to close the parentheses after typing in the radicand.

(a) Evaluate $\sqrt{10}$ and round the result to three places after the decimal.

$\boxed{\text{2nd}}\ \boxed{\sqrt{}}\ \boxed{1}\ \boxed{0}\ \boxed{)}\ \boxed{\text{ENTER}}$

The result is 3.16227766 and rounds to 3.162.

(b) Evaluate $\dfrac{\sqrt{10}}{3}$ and round the result to three places after the decimal.

2nd √ 1 0) ÷ 3 ENTER

The result rounds to 1.054.

Be careful to close the parentheses. If you do not close the parentheses, you will get the result of $\sqrt{\frac{10}{3}} \approx 1.826$.

Example Some expressions require us to use powers.

(a) Evaluate 3.2^2.

In this case, we can use the x^2 key.

3 . 2 x^2 ENTER

The result is 10.24.

(b) Evaluate 0.4^3.

In this case we use the $\wedge$ key.

. 4 $\wedge$ 3 ENTER

The result is 0.064.

ENTERING DATA

To use the statistical processes built into the TI-83 PLUS, we first enter data into lists.
Press STAT. Next we will clear any existing data lists. With EDIT highlighted, select **4:ClrList** using the arrow keys, and press ENTER.

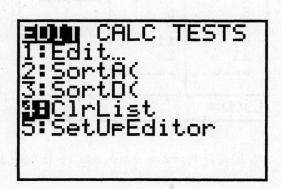

Then type in the six lists separated by commas as shown. Press ENTER.

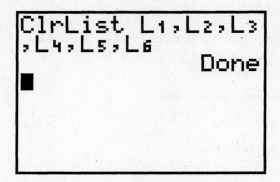

Next press STAT again and select item **1:Edit**. You will see the data screen and are now ready to enter data.

Let's enter the numbers

   2   5   7   9   11 in list L_1

   3   6   9 11   1 in list L_2

Press ENTER after each number, and use the arrow keys to move between L_1 and L_2.

To correct a data entry error, highlight the entry that is wrong and enter the correct data value.

We can also create new lists by doing arithmetic on existing lists. For instance, let's create L_3 by using the formula $L_3 = 2L_1 + L_2$.

Highlight L_3. Then type in $2L_1 + L_2$ and press ENTER.

```
L1        L2        ▓▓        3
2         3         ------
5         6
7         9
9         11
11        1
------    ------

L3 =2L₁+L₂█
```

The final result is shown.

```
L1        L2        L3        3
2         3         7
5         6         16
7         9         23
9         11        29
11        1         23
------    ------    ------

L3(1)=7
```

To leave the data screen, press 2nd [QUIT] or press another menu key, such as STAT.

COMMENTS ON ENTERING AND CORRECTING DATA

(a) To create a new list from old lists, the lists must all have the same number of entries.

(b) To delete a data entry, highlight the data value you wish to correct and press the DEL key.

(c) To insert a new entry into a list, highlight the position directly below the place you wish to insert the new value. Press 2nd [INS] and then enter the data value.

LAB ACTIVITIES TO GET STARTED USING THE TI-83 PLUS

1. Practice doing some calculations using the TI-83 Plus: Be sure to use parentheses around a numerator or denominator that contains more than a single number. Round all answers to three places after the decimal.

(a) $\dfrac{5-2.3}{1.3}$ Ans. 2.077

(b) $\dfrac{8-3.3}{\frac{3}{2}}$ Ans. 3.133

(c) $-2(3.4)+5.8$ Ans. −1

(d) $-4(-1.7)-2.1$ Ans. 4.7

(e) $\sqrt{5.3}$ Ans. 2.302

(f) $\sqrt{6+3(2)}$ Ans. 3.464

(g) $\sqrt{\dfrac{8-2.7}{5-1}}$ Ans. 1.151

(h) 1.5^2 Ans. 2.25

(i) $(5-7.2)^2$ Ans. 4.84

(j) $(0.7)^3(0.3)^2$ Ans. 0.031

2. Enter the following data into the designated list. Be sure to clear all lists first.

L_1: 3   7   9.2   12   −4

L_2: −2   9   4.3   16   10

L_3: $L_1 - 2$

L_4: $-2L_2 - L_1$

L_1: Change the data value in the second position to 6.

RANDOM SAMPLES
(SECTION 1.2 OF *UNDERSTANDING BASIC STATISTICS*)

The TI-83 Plus graphics calculator has a random number generator, which can be used in place of a random number table. Press MATH. Use the arrow keys to highlight **PRB**. Notice that **1:rand** is selected.

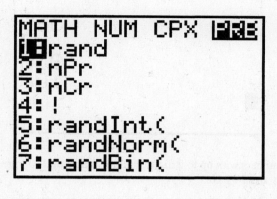

When you press ENTER, rand appears on the screen. Press ENTER again and a random number between 0 and 1 appears.

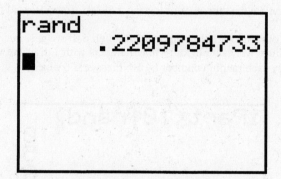

To generate a random whole number with up to 3 digits, multiply the output of **rand** by 1000 and take the integer part. The integer-part command is found in the MATH menu under **NUM**, selection **3:iPart.** Press ENTER to display the command on the main screen.

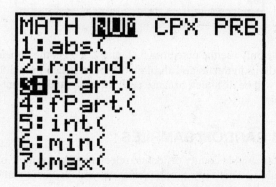

Now let's put all commands together to generate random whole numbers with up to 3 digits. First display **iPart.** Then enter **1000*rand** and close the parentheses. Now, each time you press ᴱᴺᵀᴱᴿ, a new random number will appear. Notice that the numbers might be repeated.

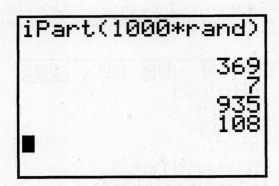

To generate random whole numbers with up to 2 digits, multiply **rand** by 100 instead of 1000. For random numbers with 1 digit, multiply **rand** by 10.

Simulating experiments in which outcomes are equally likely is an important use of random numbers.

Example

Use the TI-83 Plus to simulate the outcomes of tossing a die six times. Record the results.

In this case, the possible outcomes are the numbers 1 through 6. If we generate a random number outside of this range we will ignore it. Since we want random whole numbers with 1 digit, we will follow the method prescribed earlier and multiply each random number by 10. Press ᴱᴺᵀᴱᴿ 6 times.

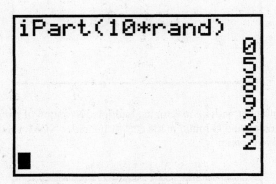

Of the first six values listed, we only use the outcomes 5, 2, and 2. To get the remaining outcomes, keep pressing ᴱᴺᵀᴱᴿ until three more digits in the range 1 through 6 appear. When we did this, the next three such digits were 1, 4, and 3. Your results will be different, because each time the random number generator is used, it gives a different sequence of numbers.

LAB ACTIVITIES FOR RANDOM SAMPLES

1. Out of a population of 800 eligible county residents, select a random sample of 20 for prospective jury duty. By what value should you multiply **rand** to generate random whole numbers with 3 digits? List the first 20 numbers corresponding to people for prospective jury duty.

2. We can stimulate dealing bridge hands by numbering the cards in a bridge deck from 1 to 52. Then we draw a random sample of 13 numbers without replacement from the population of 52 numbers. A bridge deck has 4 suits: hearts, diamonds, clubs, and spades. Each suit contains 13 cards: those numbered 2 through 10, a jack, a queen, a king, and an ace. Decide how to assign the numbers 1 through 52 to the cards in the deck. Use the random number generator on the TI-83 Plus to get the numbers of the 13 cards in one hand. Translate the numbers to specific cards and tell what cards are in the hand. For a second game, the cards would be collected and reshuffled. Using random numbers from the TI-83 Plus, determine the hand you might get in a second game.

CHAPTER 2 ORGANIZING DATA

HISTOGRAMS
(SECTION 2.2 OF *UNDERSTANDING BASIC STATISTICS*)

The TI-83 Plus graphing calculator draws histograms for data entered in the data lists. The calculator follows the convention that if a data value falls on a class limit, then it is tallied in the frequency of the bar to the right. However, if you specify the lower class boundary of the first class and the class width (see *Understanding Basic Statistics* for procedures to find these values), then no data will fall on a class boundary. The following example shows you how to draw histograms.

Example

Throughout the day from 8 A.M. to 11 P.M., Tiffany counted the number of ads occurring every hour on one commercial T.V. station. The 15 data values are

$$10 \quad 12 \quad 8 \quad 7 \quad 15 \quad 6 \quad 5 \quad 8$$
$$8 \quad 10 \quad 11 \quad 13 \quad 15 \quad 8 \quad 9$$

To make a histogram with 4 classes:

First enter the data in L_1.

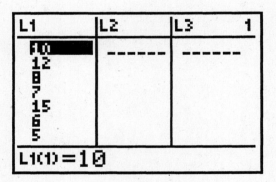

Next we will set the graphing window. However, we need to know both the lower class boundary for the first class and the class width. Use the techniques in Section 2.2 of *Understanding Basic Statistics* to find these values.

The smallest data value is 5, so the lower class boundary of the first class is 4.5.

The class width is

$$\frac{\text{largest data value} - \text{smallest data value}}{\text{Number of classes}} \text{ } increased \text{ } to \text{ } the \text{ } next \text{ } integer$$

$$\frac{15-5}{4} = 2.5 \text{ } increased \text{ } to \text{ } 3$$

Now we set the graphing window. Press the ⟨WINDOW⟩ key.

Select **Xmin** and enter the *lower class boundary* of the first class. For this example, **Xmin = 4.5**.

Select **Xscl** and enter the *class width*. For this example, **Xscl = 3**.

Setting **Xmax = 17** ensures that the histogram fits on the screen, since our highest data value is 15.

Setting **Ymin = -5** and **Ymax = 15** makes sure the histogram fits comfortably on the screen.

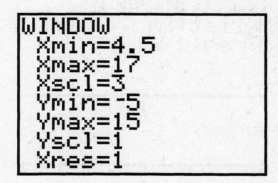

The next step is to select histogram style for the plot. Press ⟨2nd⟩ [**STAT PLOT**].

Highlight **1:Plot1** and press ⟨ENTER⟩.

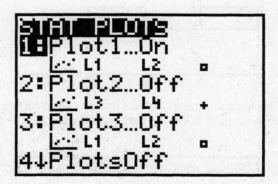

Then on the next screen select **On** and the histogram shape.

In this example we stored the data in L_1. Set Xlist to L1 by pressing [2nd] **[L1].** Each data value is listed once for each time it occurs, so **1** is entered for **Freq.**

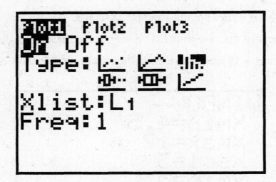

We are ready to graph the histogram. Press [GRAPH].

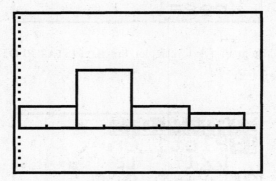

Press [TRACE] and notice that a blinking cursor appears over the first bar. The class boundaries of the bar are given as **min** and **max.** The frequency is given by **n.** Use the right and left arrow keys to move from bar to bar.

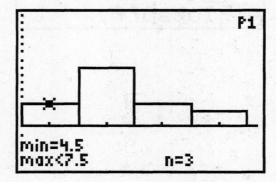

After you have graphed the histogram, you can use [TRACE] and the arrow keys to find the class boundaries and frequency of each class. The calculator has done the work of sorting the data and tallying the data for each class.

LAB ACTIVITIES FOR HISTOGRAMS

1. A random sample of 50 professional football players produced the following data.

   Weights of Pro Football Players (data file Svls02.txt)

   The following data represents weights in pounds of 50 randomly selected pro football linebackers.

   Source: The Sports Encyclopedia Pro Football 1960–1992

225	230	235	238	232	227	244	222
250	226	242	253	251	225	229	247
239	223	233	222	243	237	230	240
255	230	245	240	235	252	245	231
235	234	248	242	238	240	240	240
235	244	247	250	236	246	243	255
241	245						

   Enter this data in L_1. Scan the data for the low and high values. Compute the class width for 5 classes and find the lower class boundary for the first class. Use the TI-83 Plus to make a histogram with 5 classes. Repeat the process for 9 classes. Is the data distribution skewed or symmetric? Is the distribution shape more pronounced with 5 classes or with 9 classes?

2. Explore the other data files found in the Appendix of this *Guide*, such as

   Disney Stock Volume (data file Svls01.txt)
   Heights of Pro Basketball Players (data file Svls03.txt)
   Miles per Gallon Gasoline Consumption (data file Svls04.txt)
   Fasting Glucose Blood Tests (data file Svls05.txt)
   Number of Children in Rural Families (data file Svls06.txt)

   Select one of these files and make a histogram with 7 classes. Comment on the histogram shape.

3. (a) Consider the data

1	3	7	8	10
6	5	4	2	1
9	3	4	5	2

   Place the data in L_1. Use the TI-83 Plus to make a histogram with $n = 3$ classes. Jot down the class boundaries and frequencies so that you can compare them to part (b).

(b) Now add 20 to each data value of part (a). The results are

21	23	27	28	30
26	25	24	22	21
29	23	24	25	22

By using arithmetic in the data list screen, you can create L_2 and let $L_2 = L_1 + 20$.

Make a histogram with 3 classes. Compare the class boundaries and frequencies with those obtained in part (a). Are each of the boundary values 20 more than those of part (a)? How do the frequencies compare?

(c) Use your discoveries from part (b) to predict the class boundaries and class frequency with 3 classes for the data values below. What do you suppose the histogram will look like?

1001	1003	1007	1008	1010
1006	1005	1004	1002	1001
1009	1003	1004	1005	1002

Would it be safe to say that we simply shift the histogram of part (a) 1000 units to the right?

(d) What if we multiply each of the values of part (a) by 10? Will we effectively multiply the entries for using class boundaries by 10? To explore this question create a new list L_3 using the formula $L_3 = 10L_1$. The entries will be

10	30	70	80	100
60	50	40	20	10
90	30	40	50	20

Compare the histogram with three classes to the one from part (a). You will see that there does not seem to be an exact correspondence. To see why, look at the class width and compare it to the class width of part (a). The class width is always increased to the next integer value no matter how large the integer data values are. Consequently, the class width for the data in part (d) was increased to 31 rather than 40.

4. Histograms are not effective displays for some data. Consider the data

1	2	3	6	5	7
9	8	4	12	11	15
14	12	6	2	1	206

Use the TI-83 Plus to make a histogram with 2 classes. Then change to 3 classes, on up to 10 classes. Notice that all the histograms lump the first 17 data values into the first class, and the one data value, 206, in the last class. What feature of the data causes this phenomenon? Recall that

$$\text{Class width} = \frac{\text{largest value} - \text{smallest value}}{\text{number of class}} \textit{ increased to the next integer.}$$

How many classes would you need before you began to see the first 17 data values distributed among several classes? What would happen if you simply did not include the extreme value 206 in your histogram?

CHAPTER 3 AVERAGES AND VARIATION

ONE-VARIABLE STATISTICS
(SECTIONS 3.1 AND 3.2 OF *UNDERSTANDING BASIC STATISTICS*)

The TI-83 Plus graphing calculator supports many of the common descriptive measures for a data set. The measured supported are

Mean $\bar{x}$

Sample standard deviations S_x

Population standard deviation σ_x

Number of data values n

Median

Minimum data value

Maximum data value

Quartile 1

Quartile 3

$\sum x$

$\sum x^2$

Although the mode is not provided directly, the menu choice **SortA** sorts the data in ascending order. By scanning the sorted list, you can find the mode fairly quickly.

Example

At Lazy River College, 15 students were selected at random from a group registering on the last day of registration. The times (in hours) necessary for these students to complete registration follow:

1.7	2.1	0.8	3.5	1.5	2.6	2.1	2.8
3.1	2.1	1.3	0.5	2.1	1.5	1.9	

Use the TI-83 Plus to find the mean, sample standard deviation, and median. Use the **SortA** command to sort the data and scan for the mode, if it exists.

First enter the data into list L_1.

19

Press ⌷ again, and highlight **CALC** with item **1:1-Var Stats.**

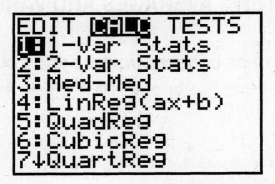

Press ⌷. The command **1-Var Stats** will appear on the screen. Then tell the calculator to use the data in list L_1 by pressing ⌷**[L₁].**

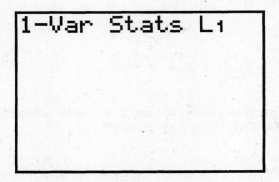

Press ⌷. The next two screens contain the statistics for the data in List L_1.

Note the arrow ↓ on the last line. This is an indication that you may use the down-arrow key to display more results.

```
1-Var Stats
 x̄=1.973333333
 Σx=29.6
 Σx²=67.68
 Sx=.813692348б
 σx=.786101491∀
↓n=15
```

Scroll down. The second full screen shows the rest of the statistics for the data in list L_1.

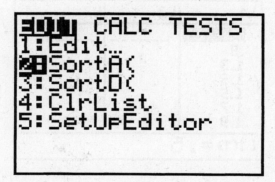

To sort the data, first press [STAT] to return to the edit menu. Then highlight **EDIT** and item **2:SortA(**.

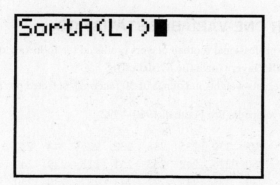

Press [ENTER]. The command **SortA** appears on the screen. Type in L_1.

When you press [ENTER], the comment **Done** appears on the screen.

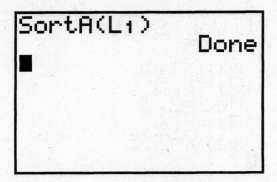

Press [STAT] and return to the data entry screen. Notice that the data in list L_1 is now sorted in ascending order.

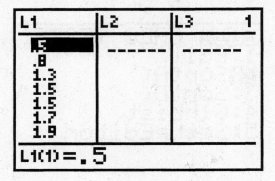

A scan of the data shows that the mode is 2.1, since this data value occurs more than any other.

LAB ACTIVITIES FOR ONE-VARIABLE STATISTICS

1. A random sample of 50 professional football players produced the following data.
   Weights of Pro Football Players (data file Svls02.txt)
   The following data represents weights in pounds of 50 randomly selected pro football linebackers.

   Source: The Sports Encyclopedia Pro Football 1960–1992

225	230	235	238	232	227	244	222
250	226	242	253	251	225	229	247
239	223	233	222	243	237	230	240
255	230	245	240	235	252	245	231
235	234	248	242	238	240	240	240
235	244	247	250	236	246	243	255
241	245						

   Enter this data in L_1. Use **1-Var Stats** to find the mean, median, and standard deviation for the weights. Use **SortA** to sort the data and scan for a mode if one exists.

2. Explore some of the other data sets found in the Appendix of this *Guide*, such as

Disney Stock Volume (data file Svls01.txt)

Heights of Pro Basketball Players (data file Svls03.txt)

Miles per Gallon Gasoline Consumption (data file Svls04.txt)

Fasting Glucose Blood Tests (data file Svls05.txt)

Number of Children in Rural Families (data file Svls06.txt)

Use the TI-83 Plus to find the mean, median, standard deviation, and mode of the data.

3. In this problem we will explore the effects of changing data values by multiplying each data value by a constant, or by adding the same constant to each data value.

(a) Consider the data

1	8	3	5	7
2	10	9	4	6
3	5	2	9	1

Enter the data into list L_1 and use the TI-83 Plus to find the mode (if it exits), mean, sample standard deviation, range, and median. Make a note of these values, since you will compare them to those obtained in parts (b) and (c).

(b) Now multiply each data value of part (a) by 10 to obtain the data

10	80	30	50	70
20	100	90	40	60
30	50	20	90	10

Remember, you can enter these data in L_2 by using the command $L_2 = 10L_1$.

Again, use the TI-83 Plus to find the mode (if it exists), mean, sample standard deviation, range, and median. Compare these results to the corresponding ones of part (a). Which values changed? Did those that changed change by a factor of 10? Did the range or standard deviation change? Referring to the formulas for these measures (see Section 3.2 of *Understanding Basic Statistics*), can you explain why the values behaved the way they did? Will these results generalize to the situation of multiplying each data entry by 12 instead of by 10? What about multiplying each by 0.5? Predict the corresponding values that would occur if we multiplied the data set of part (a) by 1000.

(c) Now suppose we add 30 to each data value of part (a)

31	38	33	35	37
32	40	39	34	36
33	35	32	39	31

To enter this data, create a new list L_3 by using the command $L_3 = L_1 + 30$.

Again use the TI-83 Plus to find the mode (if it exists), mean, sample standard deviation, range, and median. Compare these results to the corresponding ones of part (a). Which values changed? Of those that are different, did each change by being 30 more than the corresponding value of part (a)? Again look at the formulas for range and standard deviation. Can you predict the observed behavior from the formulas? Can you generalize these results? What if we added 50 to each data value of part (a)? Predict the values for the mode, mean, sample standard deviation, range, and median.

BOX-AND-WHISKER PLOTS
(SECTION 3.3 OF *UNDERSTANDING BASIC STATISTICS*)

The box-and-whisker plot is based on the five-number summary values found under **1-Var Stats:**

> Lowest value
>
> Quartile 1, or Q_1
>
> Median
>
> Quartile 3, or Q_3
>
> Highest value

Example

Let's make a box-and-whisker plot using the data about the time it takes to register for Lazy River College students who waited till the last day to register. The times (in hours) are

1.7	2.1	0.8	3.5	1.5	2.6	2.1	2.8
3.1	2.1	1.3	0.5	2.1	1.5	1.9	

In a previous example we found the **1-Var Stats** for this data. First enter the data into list L_1.

Press [WINDOW] to set the graphing window. Use Xmin = 0.5, since that is the smallest value. Use Xmax = 3.5, since that is the largest value. Use Xscl = 1. Use Ymin = –5 and Ymax = 5 to position the axis in the window.

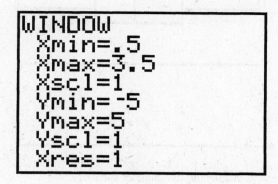

Press [2nd] [STAT PLOT] and highlight **1:Plot.** Press [ENTER].

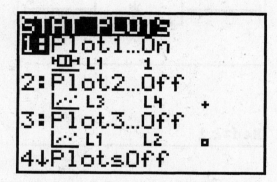

Highlight **On,** and select the box plot. The Xlist should be **L₁** and Freq should be **1.**

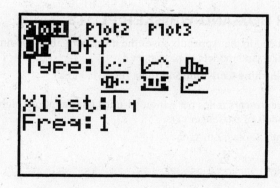

Press [GRAPH].

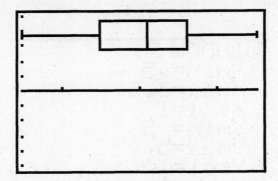

Press [TRACE] and use the arrow keys to display the five-number summary.

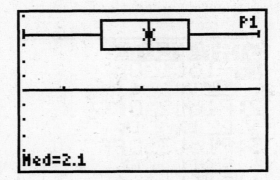

You now have many descriptive tools available: histograms, box-and-whisker plots, averages, and measures of variation such as the standard deviation. When you use all of these tools, you get a lot of information about the data.

LAB ACTIVITIES FOR BOX-AND-WHISKER PLOTS

1. One of the data sets included in the Appendix gives the miles per gallon gasoline consumption for a random sample of 55 makes and models of passenger cars.

    Miles per Gallon Gasoline Consumption (data file Svls04.txt)

    The following data represents miles per gallon gasoline consumption (highway) for a random sample of 55 makes and models of passenger cars.

    Source: Environmental Protection Agency

30	27	22	25	24	25	24	15	35	35	33
52	49	10	27	18	20	23	24	25	30	24
24	24	18	20	25	27	24	32	29	27	24
27	26	25	24	28	33	30	13	13	21	28
37	35	32	33	29	31	28	28	25	29	31

Enter these data into list L_1, and then make a histogram and a box-and-whisker plot, compute the mean, median, and sample standard deviation. Based on the information you obtain, respond to the following questions:

(a) Is the distribution skewed or symmetric? How is this shown in both the histogram and the box-and-whisker plot?

(b) Look at the box-and-whisker plot. Are the data more spread out above the median or below the median?

(c) Look at the histogram and estimate the location of the mean on the horizontal axis. Are the data more spread out above the mean or below the mean?

(d) Do there seem to be any data values that are unusually high or unusually low? If so, how do these show up on a histogram or on a box-and-whisker plot?

(e) Pretend that you are writing a brief article for a newspaper. Describe the information about the data in non-technical terms. Be sure to make some comments about the "average" of the data values and some comments about the spread of the data.

2. (a) Consider the test scores of 30 students in a political science class.

85	73	43	86	73	59	73	84	62	100
75	87	70	84	97	62	76	89	90	83
70	65	77	90	84	80	68	91	67	79

For this population of test scores find the mode, median, mean, range, variance, standard deviation, CV, the five-number summary, and make a box-and-whisker plot. Be sure to record all of these values so you can compare them to the results of part (b).

(b) Suppose Greg was in the political science class of part (a). Suppose he missed a number of classes because of illness, but took the exam anyway and made a score of 30 instead of 85 as listed as the first entry of the data in part (a). Again, use the calculator to find the mode, median, mean, range, variance, standard deviation, CV, the five-number summary, and make a box-and-whisker plot using the new data set. Compare these results to the corresponding results of part (a). Which average was most affected: mode, median, or mean? What about the range, standard deviation, and coefficient of variation? How do the box-and-whisker plots compare?

(c) Write a brief essay in which you use the results of parts (a) and (b) to predict how an extreme data value affects a data distribution. What do you predict for the results if Greg's test score had been 80 instead of 30 or 85?

CHAPTER 4 REGRESSION AND CORRELATION

LINEAR REGRESSION
(SECTION 4.1–4.3 OF *UNDERSTANDING BASIC STATISTICS*)

Important Note: Before beginning this chapter, press [2nd] **[CATALOG] (above** [0] **) and scroll down to the entry DiagnosticOn. Press** [ENTER] **twice.** After doing this, the regression correlation coefficient *r* will appear as output with the linear regression line.

The TI-83 Plus graphing calculator has automatic regression functions built-in as well as summary statistics for two variables. To locate the regression menu, press [STAT], and then select **CALC.** The choice **8:LinReg(a + bx)** performs linear regression on the variables in L_1 and L_2. If the data are in other lists, then specify those lists after the **LinReg(a + bx)** command.

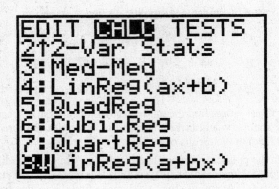

Let's look at a specific example to see how to do linear regression on the TI-83 Plus.

Example

Merchandise loss due to shoplifting, damage, and other causes is called shrinkage. Shrinkage is a major concern to retailers. The managers of H.R. Merchandise believe there is a relationship between shrinkage and number of clerks on duty. To explore this relationship, a random sample of 7 weeks was selected. During each week the staffing level of sales clerks was kept constant and the dollar value of the shrinkage was recorded.

Number of Sales Clerks:	12	11	15	9	13	8
Shrinkage:	15	20	9	25	12	31

(a) Find the equation of the least-squares line, with number of sales clerks as the explanatory variable.

First put the numbers of sales clerks in L_1 and the corresponding amount of shrinkage in L_2.

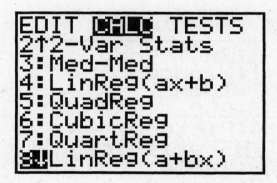

Use the **CALC** menu and select **8:LinReg(a+bx)**.

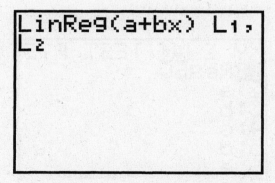

Designate the lists containing the data. Then press ENTER.

```
LinReg
 y=a+bx
 a=54.48
 b=-3.16
 r²=.9638610039
 r=-.9817642303
```

Note that the linear regression equation is $y = 54.48x - 3.16x$.

We also have the value of the Pearson correlation coefficient r and r^2.

(b) Make a scatter diagram and show the least-squares line on the plot.

To graph the least squares line, press ⎡Y=⎤. Clear all existing equations from the display. To enter the least-squares equation directly, press ⎡VARS⎤. Select item **5:Statistics.**

```
VARS Y-VARS
1:Window…
2:Zoom…
3:GDB…
4:Picture…
5 Statistics…
6:Table…
7:String…
```

Press ⎡ENTER⎤ and then select **EQ** and then **1:RegEQ.**

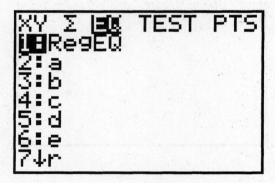

You will see the least-squares regression equation appear automatically in the equations list.

```
Plot1 Plot2 Plot3
\Y1▉54.479999999
997+ -3.159999999
9998X
\Y2=
\Y3=
\Y4=
\Y5=
```

We want to graph the least-squares line with the scatter diagram. Press [2nd] **[STAT PLOT]**. Select **1:Plot1.** Then select **On**, scatter diagram picture, and enter L_1 for Xlist and L_2 for Ylist.

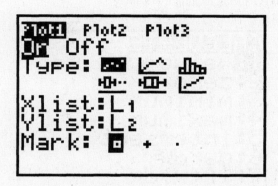

The last step before we graph is to set the graphing window. Press [WINDOW]. Set Xmin to the lowest *x* value, Xmax to the highest *x* value, ymin at –5 and ymax at above the highest *y* value.

```
WINDOW
 Xmin=8
 Xmax=15
 Xscl=1
 Ymin=-5
 Ymax=30
 Yscl=5
 Xres=1
```

Now press [GRAPH].

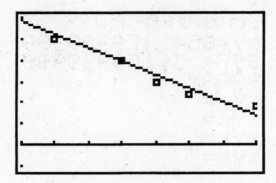

(c) Predict the shrinkage when 10 clerks are on duty.

Press [2nd] **[CALC].** This is the Calculate function for graphs and is different from the CALC you highlight from the STAT menu. Highlight **1:Value.**

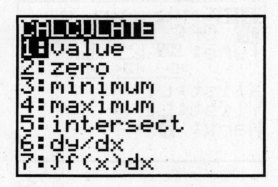

Press [ENTER]. The graph will appear. Enter 10 next to **X=.**

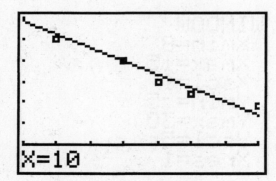

Press ⏎. The predicted value for *y* appears.

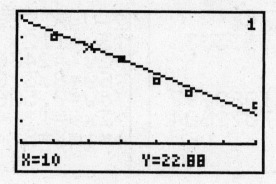

To display the summary statistics for *x* and *y*, press [STAT], highlight **CALC**. Select **2:2-Var Stats**. Press ⏎.

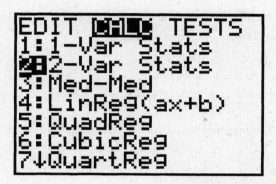

Enter the lists containing the data and press ⏎.

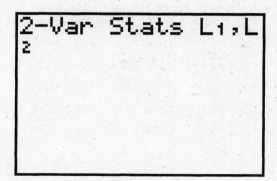

Scroll to the statistics of interest.

```
2-Var Stats
 x̄=11.33333333
 Σx=68
 Σx²=804
 Sx=2.581988897
 σx=2.357022604
↓n=6
```

```
2-Var Stats
↑Σy=112
 Σy²=2436
 Sy=8.310635756
 σy=7.586537784
 Σxy=1164
↓minX=8
```

LAB ACTIVITIES FOR LINEAR REGRESSION

For each of the following data sets, do the following.

   (a) Enter the data, putting x values in L_1 and y values in L_2.

   (b) Find the equation of the least-squares line.

   (c) Find the value of the correlation coefficient r.

   (d) Draw a scatter diagram and show the least-squares line on the scatter diagram.

1. **Cricket Chirps Versus Temperature** (data files Slr02L1.txt and Slr02L2.txt)

   In the following data pairs (X, Y)

      X = Chirps/sec for the striped ground cricket

      Y = Temperature in degrees Fahrenheit

   Source: *The Song of Insects* by Dr. G.W. Pierce, Harvard College Press

(20.0, 88.6)	(16.0, 71.6)	(19.8, 93.3)
(18.4, 84.3)	(17.1, 80.6)	(15.5, 75.2)
(14.7), 69.7)	(17.1, 82.0)	(15.4, 69.4)
(16.2, 83.3)	(15.0, 79.6)	(17.2, 82.6)
(16.0, 80.6)	(17.0, 83.5)	(14.4, 76.3)

2. List Price Versus Best Price for a New GMC Pickup Truck (data files Slr01L1.txt and Slr01L2.txt)

In the following data pairs (X, Y)

X = List price (in $1000) for a GMC Pickup Truck

Y = Best price (in $1000) for a GMC Pickup Truck

Source: Consumers Digest, February 1994

(12.400, 11.200)   (14.300, 12.500)   (14.500, 12.700)

(14.900, 13.100)   (16.100, 14.100)   (16.900, 14.800)

(16.500, 14.400)   (15.400, 13.400)   (17.000, 14.900)

(17.900, 15.600)   (18.800, 16.400)   (20.300, 17.700)

(22.400, 19.600)   (19.400, 16.900)   (15.500, 14.000)

(16.700, 14.600)   (17.300, 15.100)   (18.400, 16.100)

(19.200, 16.800)   (17.400, 15.200)   (19.500, 17.000)

(19.700, 17.200)   (21.200, 18.600)

3. Diameter of Sand Granules Versus Slope on a Natural Occurring Ocean Beach (data files Slr03L1.txt and Slr03L2.txt)

In the following data pairs (X,Y)

X = Median diameter (mm) of granules of sand

Y = Gradient of beach slope in degrees

The data is for naturally occurring ocean beaches.

Source: *Physical Geography* by A.M. King, Oxford Press, England

(0.170, 0.630)   (0.190, 0.700)   (0.220, 0.820)

(0.235, 0.880)   (0.235, 1.150)   (0.300, 1.500)

(0.350, 4.400)   (0.420, 7.300)   (0.850, 11.300)

4. National Unemployment Rate Male Versus Female (data files Slr04L1.txt and Slr04L2.txt)

In the following data pairs (X, Y)

X = National unemployment rate for adult males

Y = National unemployment rate for adult females

Source: Statistical Abstract of the United States

(2.9, 4.0)   (6.7, 7.4)   (4.9, 5.0)

(7.9, 7.2)   (9.8, 7.9)   (6.9, 6.1)

(6.1, 6.0)   (6.2, 5.8)   (6.0, 5.2)

(5.1, 4.2)   (4.7, 4.0)   (4.4, 4.4)

(5.8, 5.2)

CHAPTER 5 ELEMENTARY PROBABILITY THEORY

There are no specific TI-83 Plus activities for the basic rules of probability. However, notice that the TI-83 has menu items for factorial notation, combinations $C_{n,r}$ and permutations $P_{n,r}$. To find these functions, press [MATH] and then highlight **PRB**.

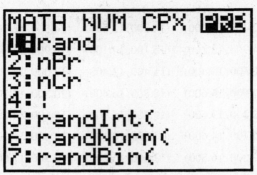

To compute $C_{12,6}$, first clear the screen and type 12. Then use the [MATH] key to select **PRB** and item **3:nCr.** Type the number 6 and then press [ENTER].

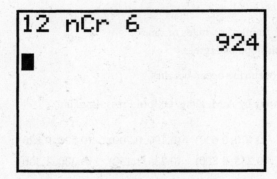

The menu option **nPr** works in a similar fashion and gives you the value of $P_{n,r}$.

CHAPTER 6 THE BINOMIAL PROBABILITY DISTRIBUTION AND RELATED TOPICS

DISCRETE PROBABILITY DISTRIBUTIONS
(SECTION 6.1 OF *UNDERSTANDING BASIC STATISTICS*)

The TI-83 Plus will compute the mean and standard deviation of a discrete probability distribution. Just enter the values of the random variable x in list L_1 and the corresponding probabilities in list L_2. Use the techniques of grouped data and one-variable statistics to compute the mean and standard deviation.

Example

How long do we spend on hold when we call a store? One study produced the following probability distribution, with the times recorded to the nearest minute.

Time on hold, x:	0	1	2	3	4	5
$P(x)$:	0.15	0.25	0.40	0.10	0.08	0.02

Find the expected value of the time on hold and the standard deviation of the probability distribution.

Enter the times in L_1 and the probabilities in list L_2. Press the STAT key, choose **CALC** and use **1-Var Stats** with lists L_1 and L_2. The results are displayed.

```
1-Var Stats
 x̄=1.77
 Σx=1.77
 Σx²=4.53
 Sx=
 σx=1.181989848
↓n=1
```

The expected value is $\bar{x} = 1.77$ minutes with standard deviation $\sigma = 1.182$.

LAB ACTIVITIES FOR DISCRETE PROBABILITY DISTRIBUTIONS

1. The probability distribution for scores on a mechanical aptitude test is

Score, x:	0	10	20	30	40	50
$P(x)$:	0.130	0.200	0.300	0.170	0.120	0.080

Use the TI-83 Plus to find the expected value and standard deviation of the probability distribution.

2. Hold Insurance has calculated the following probabilities for claims on a new $20,000 vehicle for one year if the vehicle is driven by a single male under 25.

$ Claims, x:	0	1000	5000	100,000	200,000
$P(x)$:	0.63	0.24	0.10	0.02	0.01

What is the expected value of the claim in one year? What is the standard deviation of the claim distribution? What should the annual premium be to include $400 in overhead and profit as well as the expected claim?

BINOMIAL PROBABILITIES
(SECTION 6.2 OF *UNDERSTANDING BASIC STATISTICS*)

For a binomial distribution with n trials, r successes, and the probability of success on a single trial p, the formula for the probability of r successes out of n trials is

$$P(r) = C_{n,r} \, p^r \, (1-p)^{n-r}$$

To compute probabilities for specific values of n, r, and p, we can use the TI-83 Plus graphing calculator.

Example

Consider a binomial experiment with 10 trials and probability of success on a single trial $p = 0.72$. Compute the probability of 7 successes out of n trials.

On the TI-83 Plus, the combinations function **nCr** is found in the [MATH] menu under **PRB.** To use the function, we first type the value of n, then **nCr,** and finally the value of r. Here, $n = 10$, $r = 7$, $p = 0.72$ and $(1 - p) = 0.28$. By the formula, we raise 0.72 to the power 7 and 0.28 to the power 3. The probability of 7 successes is 0.264.

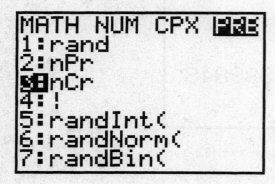
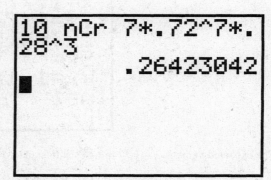

Using the Probability Distributions on the TI-83 Plus

The TI-83 Plus fully supports the binomial distribution and has it built in as a function. To access the menu that contains the probability functions, press ⌨ [DISTR]. Then scroll to item **0:binomial pdf(** and press ⌨.

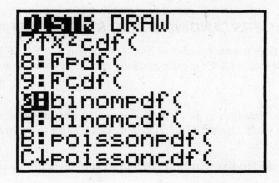

Type in the number of trials, followed by the probability of success on a single trial, followed by the number of successes. Separate each entry by a comma. Press ⌨.

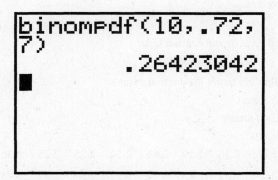

LAB ACTIVITIES FOR BINOMIAL PROBABILITIES

1. Consider a binomial distribution with $n = 8$ and $p = 0.43$. Use the formula to find the probability of r successes for r from 0 through 8.

2. Consider a binomial distribution with $n = 8$ and $p = 0.43$. Use the built-in binomial probability distribution function to find the probability of r successes for r from 0 through 8.

CHAPTER 7 NORMAL DISTRIBUTIONS

CONTROL CHARTS
(SECTION 7.1 OF *UNDERSTANDING BASIC STATISTICS*)

Although the TI-83 Plus does not have a control chart option built into **STAT PLOT,** we can use one of the features in **STAT PLOT** combined with the regular graphing options to create a control chart.

Example

Consider the data from the data files in the Appendix regarding yield of wheat at Rothamsted Experiment Station over a period of thirty years. Use the TI-83 Plus to make a control chart for this data using the target mean and standard deviation values.

Yield of Wheat at Rothamsted Expreiment Station, England (data file Tscc01.txt)

The following data represent annual yield of wheat in tonnes (one ton = 1.016 tonne) for an experimental plot of land at Rothamsted Experiment Station U.K. over a period of thirty consecutive years.

Source: Rothamsted Experiment Station U.K.

We will use the following target production values:
  target mu = 2.6 tonnes
  target sigma = 0.40 tonnes

1.73	1.66	1.36	1.19	2.66	2.14	2.25	2.25	2.36	2.82
2.61	2.51	2.61	2.75	3.49	3.22	2.37	2.52	3.43	3.47
3.20	2.72	3.02	3.03	2.36	2.83	2.76	2.07	1.63	3.02

First we enter the data by row. In list L_1 put the year numbers 1 through 30. In list L_2 put the corresponding annual yield. Again, read the data by row.

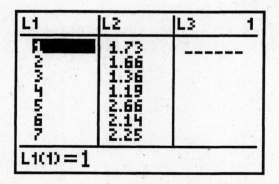

Then press [2nd] [**STAT PLOT**] and select **1:Plot1**.

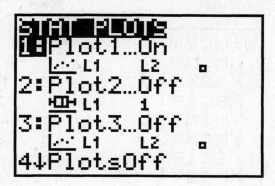

Highlight **On.** Select scatter diagram for type, L_1 for **Xlist,** and L_2 for **Ylist.** Select the symbol you like for **Mark.**

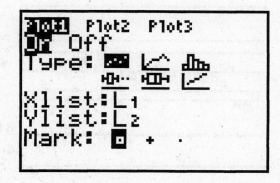

Next Press [Y=].

For Y_1 enter the target mean. In this case, enter 2.6.

For Y_2 enter the mean + 2 standard deviations. In this case, enter 2.6 + 2(.4).

For Y_3 enter the mean − 2 standard deviations. In this case, enter 2.6 − 2(.4).

For Y_4 enter the mean + 3 standard deviations. In this case, enter 2.6 + 3(.4).

For Y_5 enter the mean − 3 standard deviations. In this case, enter 2.6 − 3(.4).

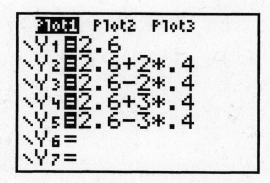

The last step is to set the graphing window. Press [WINDOW]. Since there were 30 years, set **Xmin** to 1 and **Xmax** to 30. To position the graph in the window, set **Ymin** to –1 and **Ymax** to 5.

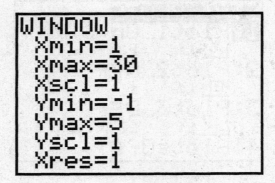

When you press [ENTER], the control chart appears.

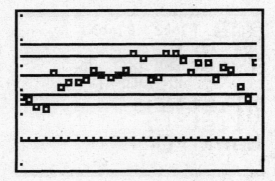

LAB ACTIVITIES FOR CONTROL CHARTS

1. Look in the Appendix and find the data file for **Futures Quotes for the Price of Coffee Beans** (Tscc04.txt). Make a control chart using the data and the target mean and target standard deviation given. Read the data by row. Are there any out-of-control signals? Explain.

2. Look in the Appendix and find the data file for **Incidence of Melanoma Tumors** (Tscc05.txt). Make a control chart using the data and the target mean and target standard deviation given. Read the data by row. Are there any out-of-control signals? Explain.

THE AREA UNDER ANY NORMAL CURVE (SECTION 7.3 OF *UNDERSTANDABLE STATISTICS*)

The TI-83 Plus gives the area under any normal distribution and shades the specified area. Press [2nd] **[DISTR]** to access the probability distributions. Select **2:normalcdf(** and press [ENTER]. Then type in the lower bound, upper bound, μ, and σ in that order separated by commas. To find the area under the normal curve with $\mu = 10$ and $\sigma = 2$ between 4 and 12, select **normalcdf(** and enter the values as shown. Then press [ENTER].

```
DISTR DRAW
1:normalpdf(
2:normalcdf(
3:invNorm(
4:tpdf(
5:tcdf(
6:X²pdf(
7↓X²cdf(
```

```
normalcdf(4,12,1
0,2)
         .8399947732
■
```

Drawing the normal distribution

To draw the graph, first set the window to accommodate the graph. Press the [WINDOW] button and enter values as shown.

```
WINDOW
 Xmin=2
 Xmax=18
 Xscl=3
 Ymin=-.1
 Ymax=.3
 Yscl=1
 Xres=1
```

Press [2nd] **[DISTR]** again and highlight **DRAW**. Select **1:ShadeNorm(**.

```
DISTR DRAW
1:ShadeNorm(
2:Shade_t(
3:ShadeX²(
4:ShadeF(
```

Again enter the lower limit, upper limit, μ, and σ separated by commas.

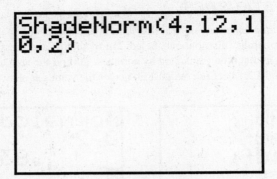

Finally press ⌷ENTER⌷.

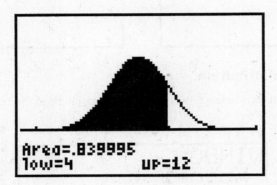

LAB ACTIVITIES FOR THE AREA UNDER ANY NORMAL CURVE

1. Find the area under a normal curve with mean 10 and standard deviation 2, between 7 and 9. Show the shaded region.

2. To find the area in the right tail of a normal distribution, select the value of 5σ for the upper limit of the region. Find the area under a normal curve with mean 10 and standard deviation 2 that lies to the right of 12.

3. Consider a random variable x that follows a normal distribution with mean 100 and standard deviation 15. Shade the regions corresponding to the probabilities and find

   (a) Shade the region corresponding to $P(x < 90)$ and find the probability.

   (b) Shade the region corresponding to $P(70 < x < 100)$ and find the probability.

   (c) Shade the region corresponding to $P(x > 115)$ and find the probability.

   (d) If the random variable were larger than 145, would that be an unusual event? Explain by computing $P(x > 145)$ and commenting on the meaning of the result.

CHAPTER 8
INTRODUCTION TO SAMPLING DISTRIBUTIONS

In this chapter, use the TI-83 Plus graphing calculator to do computations. For example, to compute the z score corresponding to a raw score from an $\bar{x}$ distribution, we use the formula

$$z = \frac{\bar{x} - \mu}{\frac{\sigma}{\sqrt{n}}}$$

To evaluate z, be sure to use parentheses as necessary.

Example

If a random sample of size 40 is taken from a distribution with mean $\mu = 10$ and standard deviation $\sigma = 2$, find the z score corresponding to $x = 9$.
We use the formula

$$z = \frac{9 - 10}{\frac{2}{\sqrt{40}}}$$

Key in the expression using parentheses: $(9 - 10) \div (2 \div \sqrt{\ } (40)$ [ENTER].

```
(9-10)/(2/√(40))
            -3.16227766
```

The result rounds to $z = -3.16$.

CHAPTER 9 ESTIMATION

CONFIDENCE INTERVALS FOR A POPULATION MEAN
(SECTION 9.1 AND 9.2 OF *UNDERSTANDING BASIC STATISTICS*)

The TI-83 Plus fully supports confidence intervals. To access the confidence interval choices, press [STAT] and select **TESTS.** The confidence interval choices are found in items 7 through B,

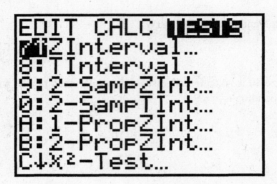

Example (Large Sample)

Suppose a random sample of 250 credit card bills showed an average balance of $1200 with a standard deviation of $350. Find a 95% confidence interval for the population mean credit card balance.

Since we have a large sample, we will use the normal distribution. Select item **7:ZInterval.**

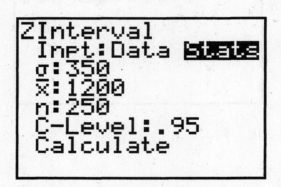

In this example, we have summary statistics, so we will select the STATS option for input. We estimate σ by the sample standard deviation 350. Then enter the value of $\overline{x}$ and the sample size n. Use 0.95 for the C-Level.

Highlight **Calculate** and press ENTER to get the results. Notice that the interval is given using standard mathematical notation for an interval. The interval for μ goes from \$1156.6 to \$1232.4.

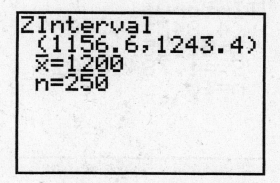

Example (Small Sample)

A random sample of 16 wolf dens showed the number of pups in each to be

5   8   7   5   3   4   3   9

5   8   5   6   5   6   4   7

Find a 90% confidence interval for the population mean number of pups in such dens.

In this case we have raw data, so enter the data in list L_1 using the **EDIT** option of the STAT key. Since we have a small sample, we use the t distribution. Under **Tests** from the STAT menu, select item **8:TInterval.** Since we have raw data, select the DATA option for Input. The data is in list L_1, and occurs with frequency 1. Enter 0.90 for the C-Level.

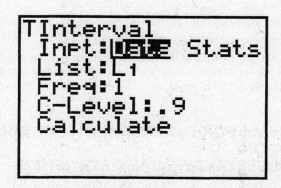

Highlight Calculate and press ⌈ENTER⌉. The result is the interval from 4.84 pups to 6.41 pups.

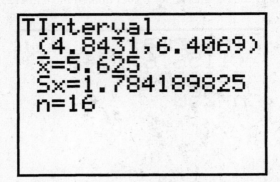

LAB ACTIVITIES FOR CONFIDENCE INTERVALS FOR A POPULATION MEAN

1. Markey Survey was hired to do a study for a new soft drink, Refresh. A random sample of 20 people were given a can of Refresh and asked to rate it for taste on a scale of 1 to 10 (with 10 being the highest rating). The ratings were

5	8	3	7	5	9	10	6	6	2
9	2	1	8	10	2	5	1	4	7

   Find an 85% confidence interval for the population mean rating of Refresh.

2. Suppose a random sample of 50 basketball players showed the average height to be 78 inches with sample standard deviation 1.5 inches.

   (a) Find a 99% confidence interval for the population mean height.

   (b) Find a 95% confidence interval for the population mean height.

   (c) Find a 90% confidence interval for the population mean height.

   (d) Find a 85% confidence interval for the population mean height.

   (e) What do you notice about the length of the confidence interval as the confidence level goes down? If you used a confidence level of 80%, would you expect the confidence interval to be longer or shorter than that of 85%? Run the program again to verify your answer.

CONFIDENCE INTERVALS FOR THE PROBABILITY OF SUCCESS *p* IN A BINOMIAL DISTRIBUTION
(SECTION 9.3 OF *UNDERSTANDING BASIC STATISTICS*)

To find a confidence interval for a proportion, press ⌈STAT⌉ and use option **A:1-PropZInt** under **TESTS**. Notice that the normal distribution will be used.

Example:

The public television station BPBS wants to find the percent of its viewing population who give donations to the station. 300 randomly selected viewers were surveyed, and it was found that 123 made contributions to the station. Find a 95% confidence interval for the probability that a viewer of BPBS selected at random contributes to the station.

The letter x is used to count the number of successes (the letter r is used in the text). Enter 123 for x and 300 for n. Use 0.95 for the C-level.

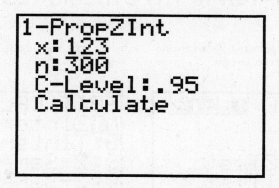

```
1-PropZInt
 x:123
 n:300
 C-Level:.95
 Calculate
```

Highlight **Calculate** and press [ENTER]. The result is the interval from 0.35 to 0.47.

```
1-PropZInt
 (.35434,.46566)
 p̂=.41
 n=300
```

LAB ACTIVITIES FOR CONFIDENCE INTERVALS FOR THE PROBABILITY OF SUCCESS p IN A BINOMIAL DISTRIBUTION

1. Many types of error will cause a computer program to terminate or give incorrect results. One type of error is punctuation. For instance, if a comma is inserted in the wrong place, the program might not run. A study of programs written by students in a beginning programming course showed that 75 out of 300 errors selected at random were punctuation errors. Find a 99% confidence interval for the proportion of errors made by beginning programming students that are punctuation errors. Next, find a 90% confidence interval. Is this interval longer or shorter?

2. Sam decided to do a statistics project to determine a 90% confidence interval for the probability that a student at West Plains College eats lunch in the school cafeteria. He surveyed a random sample of 12 students and found that 9 ate lunch in the cafeteria. Can Sam use the program to find a confidence interval for the population proportion of students eating in the cafeteria? Why or why not? Try **1-PropZInt** with $n = 12$ and $r = 9$. What happens? What should Sam do to complete his project?

CHAPTER 10 HYPOTHESIS TESTING

The TI-83 Plus fully supports hypothesis testing. Use the [STAT] key, then highlight **TESTS.** The options used in Chapter 9 are given on the two screens.

TESTING A SINGLE POPULATION MEAN
(SECTION 10.1–10.4 OF *UNDERSTANDING BASIC STATISTICS*)

For large-sample test of a mean, we use the normal distribution. Select option **1:Z-Test.** For large samples we may estimate the population standard deviation σ by the sample standard deviation s. As with confidence intervals, we have a choice of entering raw data into a list using the **Data** input option or using summary statistics with the **Stats** option. The null hypothesis is $H_0: \mu = \mu_0$. Enter the value of μ_0 in the spot indicated. To select the alternate hypothesis, use one of the three μ: options $\neq \mu_0, < \mu_0,$ or $> \mu_0$. Output consists of the value of the sample mean $\bar{x}$ and its corresponding z value. The P value of the sample statistic is also given.

Example (Testing a mean, large sample)

Ten years ago, State College did a study regarding the number of hours full-time students worked each week. The mean number of hours was 8.7. A recent study involving a random sample of 45 full time students showed that the average number of hours worked per week was 10.3 with standard deviation 2.8. Use a 5% level of confidence to test if the mean number of hours worked per week by full-time students has increased.

Because we have a large sample, we will use the normal distribution for our sample test statistic. Select option **1:Z-Test.** We have summary statistics rather than raw data, so use the **Stats** input option. Enter the appropriate values on the screen.

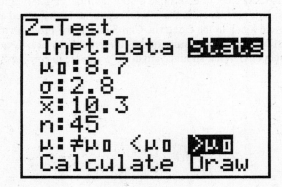

Next highlight **Calculate** and press ⏎.

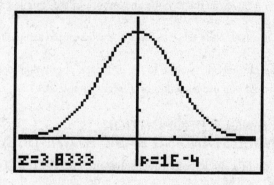

```
Z-Test
 μ>8.7
 z=3.83325939
 P=6.3250806E-5
 x̄=10.3
 n=45
```

We see that the z value corresponding to the sample test statistic $\bar{x}$ = 10.3 is z = 3.83. The critical value for a 5% level of significance and a right-tailed test is z_0 = 1.645. Clearly the sample z value is to the right of z_0, and we reject H_0. Notice that the P value = 6.325E-5. This means that we move the decimal 5 places to the left, giving a P value of 0.000063. Since the P value is less than 0.05, we reject H_0.

The TI-83 Plus gives an option to show the sample test statistic on the normal distribution. Highlight the **Draw** option on the **Z-Test** screen. Because the sample z is so far to the right, it does not appear on this window. However, its value does, and a rounded P value shows as well.

```
z=3.8333          P=1E-4
```

To do hypothesis testing of the mean for small samples, we use Student's t distribution. Select option **2:T-Test.** The entry screens are similar to those for the Z-Test.

LAB ACTIVITIES FOR TESTING A SINGLE POPULATION MEAN

1. A random sample of 65 pro basketball players showed their heights (in feet) to be

6.50	6.25	6.33	6.50	6.42	6.67	6.83	6.82
6.17	7.00	5.67	6.50	6.75	6.54	6.42	6.58
6.00	6.75	7.00	6.58	6.29	7.00	6.92	6.42
5.92	6.08	7.00	6.17	6.92	7.00	5.92	6.42
6.00	6.25	6.75	6.17	6.75	6.58	6.58	6.46
5.92	6.58	6.13	6.50	6.58	6.63	6.75	6.25
6.67	6.17	6.17	6.25	6.00	6.75	6.17	6.83
6.00	6.42	6.92	6.50	6.33	6.92	6.67	6.33
6.08							

(a) Enter the data into list L_1. Use **1-Var Stats** to determine the sample standard deviation.

(b) Use the **Z-Test** option to test the hypothesis that the average height of the players is greater than 6.2 feet, at the 1% level of significance.

2. In this problem we will see how the test conclusion is possibly affected by a change in the level of significance.

Teachers for Excellence is interested in the attention span of students in grades 1 and 2, now as compared to 20 years ago. They believe it has decreased. Studies done 20 years ago indicate that the attention span of children in grades 1 and 2 was 15 minutes. A study sponsored by Teachers for Excellence involved a random sample of 20 students in grades 1 and 2. The average attention span of these students was (in minutes) $\bar{x} = 14.2$ with standard deviation $s = 1.5$.

(a) Conduct the hypothesis test using $\alpha = 0.05$ and a left-tailed test. What is the test conclusion? What is the P value?

(b) Conduct the hypothesis test using $\alpha = 0.01$ and a left-tailed test. What is the test conclusion? How could you have predicted this result by looking at the P value from part (a)? Is the P value for this part the same as it was for part (a)?

3. In this problem, let's explore the effect that sample size has on the process of testing a mean. Run **Z-Test** with the hypotheses $H_0: \mu = 200$, $H_1: \mu > 200$, $\alpha = 0.05$, $\bar{x} = 210$ and $s = 40$.

(a) Use the sample size $n = 30$. Note the P value and z score of the sample test statistic and test conclusion.

(b) Use the sample size $n = 50$. Note the P value and z score of the sample test statistic and test conclusion.

(c) Use the sample size $n = 100$. Note the P value and z score of the sample test statistic and test conclusion.

(d) In general, if your sample statistic is close to the proposed population mean specified in H_0, and you want to reject H_0, would you use a smaller or a larger sample size?

TESTS INVOLVING A SINGLE PROPORTION (SECTION 10.5 OF *UNDERSTANDING BASIC STATISTICS*)

To conduct a hypotheses test of a single proportion, select option **5:1-PropZTest.** The null hypothesis is $H_0: p = p_0$. Enter the value of p_0. The number of successes is designated by the value x. Enter the value. The sample size or number of trials is n. The alternate hypothesis will be prop $\neq p_0$, $< p_0$, or $> p_0$. Highlight the appropriate choice. Finally highlight **Calculate** and press [ENTER]. Notice that the **Draw** option is available to show the results on the standard normal distribution.

CHAPTER 11 INFERENCES ABOUT DIFFERENCES

TESTS INVOLVING PAIRED DIFFERENCES (DEPENDENT SAMPLES) (SECTION 11.1 OF *UNDERSTANDING BASIC STATISTICS*)

To perform a paired difference test, we put our paired data into two columns, and then put the differences between corresponding pairs of values in a third column. For example, put the "before" data in list L_1, the "after" data in L_2. Create $L_3 = L_1 - L_2$.

Example

Promoters of a state lottery decided to advertise the lottery heavily on television for one week during the middle of one of the lottery games. To see if the advertising improved ticket sales, the promoters surveyed a random sample of 8 ticket outlets and recorded weekly sales for one week before the television campaign and for one week after the campaign. The results follow (in ticket sales) where B stands for "before" and A for "after" the advertising campaign.

B:	3201	4529	1425	1272	1784	1733	2563	3129
A:	3762	4851	1202	1131	2172	1802	2492	3151

Test the claim that the television campaign increased lottery ticket sales at the 0.05 level of significance.

We want to test to see if $D = B - A$ is less than zero, since we are testing the claim that the lottery ticket sales are greater after the television campaign. We will put the "before" data in L_1, the "after" data in L_2, and put the differences in list L_3 by highlighting the header L£ entering $L_1 - L_2$, and pressing ENTER.

L1	L2	**L3**	3
3201	3762	-561	
4529	4851	-322	
1425	1202	223	
1272	1131	141	
1784	2172	-388	
1733	1802	-69	
2563	2492	71	

L3 = L₁−L₂

Next we will conduct a *t*-test on the differences in list L_1. Select option **2:T-Test** from the **TESTS** menu. Select **Data** for **Inpt.** Since the null hypotheses is that $d = 0$, we use the null hypothesis $H_0: d = \mu_0$ with $\mu_0 = 0$. Since the data is in L_3, enter that as the List. Since the test is a left-tailed test, select $\mu: < \mu_0$.

Highlight **Calculate** and press ⏎.

```
T-Test
 µ<0
t=-1.178455137
P=.138559057
x̄=-115.875
Sx=278.1132542
n=8
```

We see that the sample *t* value is –1.17 with a corresponding *P* value of 0.138. Since the *P* value is greater than 0.05, we do not reject H_0.

LAB ACTIVITIES USING TESTS INVOLVING PAIRED DIFFERENCES (DEPENDENT SAMPLES)

1. The data are pairs of values where the first entry represents average salary ($1000/yr) for male faculty members at an institution and the second entry represents the average salary for female faculty members ($1000/yr) at the same institution. A random sample of 22 U.S. colleges and universities was used (source: *Academe, Bulletin of the American Association of University Professors*).

(34.5, 33.9)	(30.5, 31.2)	(35.1, 35.0)	(35.7, 34.2)	(31.5, 32.4)
(34.4, 34.1)	(32.1, 32.7)	(30.7, 29.9)	(33.7, 31.2)	(35.3, 35.5)
(30.7, 30.2)	(34.2, 34.8)	(39.6, 38.7)	(30.5, 30.0)	(33.8, 33.8)
(31.7, 32.4)	(32.8, 31.7)	(38.5, 38.9)	(40.5, 41.2)	(25.3, 25.5)
(28.6, 28.0)	(35.8, 35.1)			

(a) Put the first entries in L_1, the second in L_2, and create L_3 to be the difference $L_1 - L_2$.

(b) Use the **T-Test** option to test the hypothesis that there is a difference in salary. What is the P value of the sample test statistic? Do we reject or fail to reject the null hypothesis at the 5% level of significance? What about at the 1% level of significance?

(c) Use the **T-Test** option to test the hypothesis that female faculty members have a lower average salary than male faculty members. What is the test conclusion at the 5% level of significance? At the 1% level of significance.

2. An audiologist is conducting a study on noise and stress. Twelve subjects selected at random were given a stress test in a room that was quiet. Then the same subjects were given another stress test, this time in a room with high-pitched background noise. The results of the stress tests were scores 1 through 20, with 20 indicating the greatest stress. The results, where B represents the score of the test administered in the quiet room and A represents the scores of the test administered in the room with the high-pitched background noise, are shown below.

Subject	1	2	4	5	6	7	8	9	10	11	12
B	13	12	16	19	7	13	9	15	17	6	14
A	18	15	14	18	10	12	11	14	17	8	16

Test the hypothesis that the stress level was greater during exposure to noise. Look at the P value. Should you reject the null hypotheses at the 1% level of significance? At the 5% level?

TESTS OF DIFFERENCE OF MEANS (INDEPENDENT SAMPLES)
(SECTION 11.2 AND 11.3 OF *UNDERSTANDING BASIC STATISTICS*)

We consider the $\bar{x}_1 - \bar{x}_2$ distribution. The null hypothesis is that there is no difference between means so $H_0: \mu_1 = \mu_2$, or $H_0: \mu_1 - \mu_2 = 0$.

Example

Sellers of microwave French fry cookers claim that their process saves cooking time. McDougle Fast Food Chain is considering the purchase of these new cookers, but wants to test the claim. Six batches of French fries cooked in the traditional way. These times (in minutes) are

$$15 \qquad 17 \qquad 14 \qquad 15 \qquad 16 \qquad 13$$

Six batches of French fries of the same weight were cooked using the new microwave cooker. These cooking times (in minutes) are

$$11 \qquad 14 \qquad 12 \qquad 10 \qquad 11 \qquad 15$$

Test the claim that the microwave process takes less time. Use $\alpha = 0.05$.

Put the data for traditional cooking in list L_1 and the data for the new method in List L_2. Since we have small samples, we want to use the Student's t distribution. Select option **4:2-SampTTest**. Select **Data** for **Inpt,** use the alternate hypothesis $\mu_1 > \mu_2$, select **Yes** for **Pooled:**

The output is on two screens.

```
2-SampTTest
 μ1>μ2
 t=2.890086704
 P=.0080523896
 df=10
 x̄1=15
↓x̄2=12.16666667
```

```
2-SampTTest
 μ1>μ2
↑Sx1=1.41421356
 Sx2=1.94079022
 SxP=1.69803808
 n1=6
 n2=6
```

Since the P value is 0.008, which is less than 0.05, we reject H_0 and conclude that the new method cooks food faster.

CONFIDENCE INTERVALS FOR $\mu_1 - \mu_2$ (INDEPENDENT SAMPLES) (SECTIONS 11.2 AND 11.3 OF *UNDERSTANDING BASIC STATISTICS*)

For tests of difference of means, press [STAT], highlight **TESTS** and use the option **9:2-SampZInt** for confidence intervals for the difference of means, large samples. Use option **0:2-SampTInt** for confidence intervals for the difference of means, small samples.

For the difference of means, you have the option of using either summary statistics or raw data in lists as input. **Data** allows you to use raw data from a list, while **Stats** lets you use summary statistics.

When finding confidence intervals for differences of means, large or small samples, use sample estimates s_1 and s_2 for corresponding population standard deviations σ_1 and σ_2. For confidence intervals for difference of means using small samples, be sure to choose **YES** for Pooled.

Example (Difference of means, large sample)

A random sample of 45 pro football players produced a sample mean height of 6.18 feet with standard deviation 0.37 feet. A random sample of 40 pro basketball players produced a mean height of 6.45 feet with standard deviation 0.31. Find a 90% confidence interval for the difference of population heights.

We select option **9:2-SampZInt,** since we have large samples and want to use the normal distribution. Since we have summary statistics, select **Inpt: Stats.** Then enter the specified values. We will use the sample standard deviations as estimates for the respective population standard deviations. The data entry requires two screens.

```
2-SampZInt
 Inpt:Data Stats
 σ1:.37
 σ2:.31
 x̄1:6.18
 n1:45
 x̄2:6.45
↓n2:40
```

```
2-SampZInt
↑σ2:.31
 x̄1:6.18
 n1:45
 x̄2:6.45
 n2:40
 C-Level:.9
 Calculate
```

Highlight **Calculate** and press ENTER.

```
2-SampZInt
(-.3914,-.1486)
x̄1=6.18
x̄2=6.45
n1=45
n2=40
```

The interval is from −0.39 to −0.14. Since zero lies outside the interval, we conclude that at the 90% level, basketball players, mean height is different from that of football players.

LAB ACTIVITIES FOR TESTING DIFFERENCE OF TWO MEANS AND FOR CONFIDENCE INTERVALS

1. Calm Cough Medicine is testing a new ingredient to see if its addition will lengthen the effective cough relief time of a single dose. A random sample of 15 doses of the standard medicine were tested, and the effective relief times were (in minutes):

42	35	40	32	30	26	51	39	33	28
37	22	36	33	41					

A random sample of 20 doses were tested when the new ingredient was added. The effective relief times were (in minutes):

43	51	35	49	32	29	42	38	45	74
31	31	46	36	33	45	30	32	41	25

Assume that the standard deviations of the relief times are equal for the two populations.

(a) Test the claim that the effective relief time is longer when the new ingredient is added. Use $\alpha = 0.01$.

(b) Find a 99% confidence interval for the difference. Are the values all positive, all negative, or mixed? What does this tell you about the comparative lengths of relief time for the two medications?

2. The following data is based on random samples of red foxes in two regions of Germany. The number of cases of rabies was counted for a random sample of 16 areas in each of two regions.

Region 1:	10	2	2	5	3	4	3	3	4	0	2	6	4	8	7	4
Region 2:	1	1	2	1	3	9	2	2	3	4	3	2	2	0	0	2

(a) Use a confidence level $c = 90\%$. Does the interval indicate that the population mean number of rabies cases in Region 1 is greater than the population mean number of rabies cases in Region 2 at the 90% level? Why or why not?

(b) Use a confidence level $c = 95\%$. Does the interval indicate that the population mean number of rabies cases in Region 1 is greater than the population mean number of rabies cases in Region 2 at the 95% level? Why or why not?

(c) Determine whether the population mean number of rabies cases in Region 1 is greater than the population mean number of rabies cases in Region 2 at the 99% level. Explain your answer. Verify your answer by running **2-SampTInt** with a 99% confidence level.

TESTING A DIFFERENCE OF PROPORTIONS (SECTION 11.4 OF *UNDERSTANDING BASIC STATISTICS*)

To conduct a hypothesis test for a difference of proportions, select option **6:2-PropZTest** and enter appropriate values. The number of successes for population 1 is designated by x1, while the number of trials is designated by n1. Likewise, the number of successes for population 2 is designated by x2 and the number of trials is designated by n2. The null hypothesis is assumed to be $H_0 : p_1 = p_2$. On the calculator screen highlight the appropriate alternate hypothesis. You again have two choices: Draw or Calculate. Calculate gives the summary statistics, the z value for the sample proportion difference, and the P value of the sample proportion difference. Draw shows the P value on the normal curve and gives the z value and P value of the sample test statistic.

To conclude the test, compare the P value with the level of significance α of your test.

CONFIDENCE INTERVALS FOR $p_1 - p_2$ (LARGE SAMPLES) (SECTION 11.4 OF *UNDERSTANDING BASIC STATISTICS*)

Press STAT, highlight TESTS and use the **B:2-PropZInt** to generate confidence intervals for the difference of two proportions. Data entry is similar to that for testing the difference of two proportions.

LAB ACTIVITIES FOR TESTING DIFFERENCE OF TWO PROPORTIONS AND FOR CONFIDENCE INTERVALS

1. A random sample of 30 police officers working the night shift showed that 23 used at least 5 sick leave days per year. Another random sample of 45 police officers working the day shift showed that 26 used at least 5 sick leave days per year. Find the 90% confidence interval for the difference of population proportions of police officers working the two shifts using at least 5 sick leave days per year. At the 90% level, does it seem that the proportions are different? Does there seem to be a difference in proportions at the 99% level? Why or why not?

2. Publisher's Survey did a study to see if the proportion of men who read mysteries is different from the proportion of women who read them. A random sample of 402 women showed that 112 read mysteries regularly (at least 6 books per year). A random sample of 365 men showed that 92 read mysteries regularly. Is the proportion of mystery readers different between men and women? Use a 1% level of significance.

   (a) Find the P value of the test conclusion.

   (b) Test the hypothesis that the proportion of women who read mysteries is *greater* than the proportion of men. Use a 1% level of significance. Is the P value for a right-tailed test half that of a two-tailed test? If you know the P value for a two-tailed test, can you draw conclusions for a one-tailed test?

CHAPTER 12 ADDITIONAL TOPICS USING INFERENCES

CHI-SQUARE TEST OF INDEPENDENCE
(SECTION 12.1 OF *UNDERSTANDING BASIC STATISTICS*)

The TI-83 Plus calculator supports tests for independence. Press $\boxed{\text{STAT}}$, highlight TESTS, and select option C: χ^2 –Test. The original observed values need to be entered in a matrix.

Example

A computer programming aptitude test has been developed for high school seniors. The test designers claim that scores on the test are independent of the type of school the student attends: rural, suburban, urban. A study involving a random sample of students from each of these types of institutions yielded the following information, where aptitude scores range from 200 to 500 with 500 indicating the greatest aptitude and 200 the least. The entry in each cell is the observed number of students achieving the indicated score on the test.

School Type

Score	Rural	Suburban	Urban
200–299	33	65	82
300–399	45	79	95
400–500	21	47	63

Using the option **C:** χ^2 **-Test...,** test the claim that the aptitude test scores are independent of the type of school attended at the 0.05 level of significance.

We need to use a matrix to enter the data. Press $\boxed{\text{2nd}}$ **[MATRX].** Highlight **EDIT,** and select **1:[A].** Press $\boxed{\text{ENTER}}$.

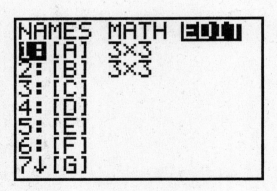

Since the contingency table has 3 rows and 3 columns, type 3 (for number of rows), press [ENTER], type 3 (for number of columns), press [ENTER]. Then type in the observed values. Press [ENTER] after each entry. Notice that you enter the table by rows. When the table is entered completely, press [2nd] [MATRX] again.

This time, select **2:[B].** Set the dimensions to be the same as for matrix [A]. For this example, [B] should be 3 rows × 3 columns.

Now press [STAT], highlight TESTS, and use option **C: χ^2 –Test.** This tells you that the observed values of the table are in matrix [A]. The expected values will be placed in matrix [B].

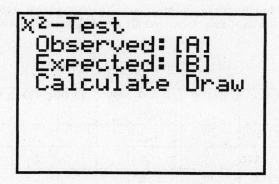

Highlight **Calculate** and press [ENTER]. We see that the sample value of χ^2 is 1.32. The P value is 0.858. Since the P value is larger than 0.05, we do not reject the null hypothesis.

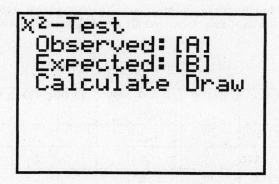

If you want to see the expected values in matrix [B], type [2nd] [MATRX] and select [B] under **NAMES.**

Graph

Notice that one of the options of the χ^2 test is to graph the χ^2 distribution and show the sample test statistic on the graph. Highlight **Draw** and press ᴱᴺᵀᴱᴿ. Before you do this, be sure that all STAT PLOTS are set to Off, and that you have cleared all the entries in the Y= menu.

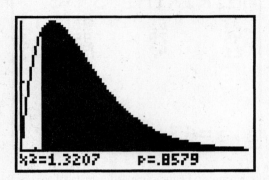

LAB ACTIVITIES FOR CHI-SQUARE TEST OF INDEPENDENCE

1. We Care Auto Insurance had its staff of actuaries conduct a study to see if vehicle type and loss claim are independent. A random sample of auto claims over six months gives the information in the contingency table.

Total Loss Claims per Year per Vehicle

Type of vehicle	$0–999	$1000–2999	$3000–5999	$6000+
Sports car	20	10	16	18
Truck	16	25	33	9
Family Sedan	40	68	17	7
Compact	52	73	48	12

Test the claim that car type and loss claim are independent. Use $\alpha = 0.05$.

2. An educational specialist is interested in comparing three methods of instruction:

   SL – standard lecture with discussion
   TV – video taped lectures with no discussion
   IM – individualized method with reading assignments and tutoring, but no lectures.

The specialist conducted a study of these three methods to see if they are independent. A course was taught using each of the three methods and a standard final exam was given at the end. Students were put into the different method sections at random. The course type and test results are shown in the next contingency table.

Final Exam Score

Course Type	< 60	60–69	70–79	80–89	90–100
SL	10	4	70	31	25
TV	8	3	62	27	23
IM	7	2	58	25	22

Test the claim that the instruction method and final test scores are independent, using $\alpha = 0.01$.

INFERENCES RELATING TO LINEAR REGRESSION (SECTIONS 12.4 AND 12.5 OF *UNDERSTANDING BASIC STATISTICS*)

The TI-83 supports testing rho, the population linear regression correlation coefficient. Press [STAT] and highlight TESTS. Then select **E:LinRegTTest**. Enter the list positions of the data (x, y). The null hypothesis is assumed to be $H_0 : \rho = 0$. Select the alternate hypothesis. The output gives the t-value of the sample correlation coefficient r. Note that in the text, the sample r-value is compared to a critical r value to conclude the test. However, there is a formula to convert the sample r value to a student's t value with degrees of freedom d.f. $= n - 2$. This is the procedure used on the TI-83. The sample t-value of the sample r value is given. In addition, the P value of the sample r value is provided. Use the P value to conclude the test by comparing it to your given level of significance α.

Other items in the output include the values of a and b for the linear regression equation $y = a + bx$. The value of the standard error of estimate S_e is given as s. In addition, the sample correlation coefficient r and the coefficient of determination r^2 are also given.

LAB ACTIVITIES FOR INFERENCES RELATING TO LINEAR REGRESSION

1. **Cricket Chirps Versus Temperature** (data files Slr02L1.txt and Slr02L2.txt)

   In the following data pairs (X, Y)

   X = Chirps/sec for the striped ground cricket

   Y = Temperature in degrees Fahrenheit

   Source: *The Song of Insects* by Dr. G.W. Pierce, Harvard College Press

(20.0, 88.6)	(16.0, 71.6)	(19.8, 93.3)
(18.4, 84.3)	(17.1, 80.6)	(15.5, 75.2)
(14.7), 69.7)	(17.1, 82.0)	(15.4, 69.4)
(16.2, 83.3)	(15.0, 79.6)	(17.2, 82.6)
(16.0, 80.6)	(17.0, 83.5)	(14.4, 76.3)

   Find the equation of the least-squares line and the value of r. Test to see if there is a positive correlation between the rate of cricket chirps and the temperature.

2. **Diameter of Sand Granules Versus Slope on a Natural Occurring Ocean Beach** (data files Slr03L1.txt and Slr03L2.txt)

In the following data pairs (X,Y)

   X = Median diameter (mm) of granules of sand

   Y = Gradient of beach slope in degrees

The data is for naturally occurring ocean beaches.

Source: *Physical Geography* by A.M. King, Oxford Press, England

(0.170, 0.630)	(0.190, 0.700)	(0.220, 0.820)
(0.235, 0.880)	(0.235, 1.150)	(0.300, 1.500)
(0.350, 4.400)	(0.420, 7.300)	(0.850, 11.300)

Find the equation of the least-squares line and the value of r. Test to see if there is any correlation between the size of sand granules and the slope of the beach.

PART II

MINITAB Guide

for

Understanding Basic Statistics, Third Edition

CHAPTER 1 GETTING STARTED

GETTING STARTED WITH MINITAB

In this chapter you will find

(a) general information about MINITAB

(b) general directions for using the Windows style pull-down menus

(c) general instructions for choosing values for dialog boxes

(d) how to enter data

(e) other general commands.

General Information

MINITAB is a command driven software package with more than 200 commands available. However, in Windows versions of MINITAB, menu options and dialog boxes can be used to generate the appropriate commands. After using the menu options and dialog boxes, the actual commands are shown in the Session window along with the output of the desired process. Data are stored and processed in a table with rows and columns. Such a table is similar to a spreadsheet and is called a worksheet. Unlike electronic spreadsheets, a MINITAB worksheet can contain only numbers and text, not formulas. Constraints are also stored in the worksheet, but are not visible.

MINITAB will accept words typed in upper or lower case letters as well as a combination of the two. Comments elaborating on the commands may be included. In this *Guide* we will follow the convention of typing the essential parts of a command in upper case letters and optional comments in lower case letters:

COMMAND with comments

Note that only the *first four letters* of a command are essential. However, we usually give the entire command name in examples.

Numbers must be typed *without* commas. Be sure you type the number zero instead of the letter "O" and that you use the number one (1) instead of a lower case letter "l." Exponential notation is also acceptable. For instance

127.5   1.257E2   1.257E+2

are all acceptable in MINITAB and have the same value.

The MINITAB *worksheet* contains columns, rows, and constants. The rows are designated by numbers. The columns are designated by the letter **C** followed by a number. **C1, C2, C3** designate columns 1, 2, and 3. Constants require the letter **K**, and may be followed by a number if there are several constants. **K1, K2** designate constant 1 and constant 2, respectively.

Starting and Ending MINITAB

The steps you use to start MINITAB will differ according to the computer equipment you are using. You will need to get specific instructions for your installation from your professor or computer lab manager. Use this space to record the details of logging onto your system and accessing MINITAB. For Windows versions, you generally click on the MINITAB icon to begin the program.

_____

_____

_____

_____

_____

_____

The first screen will look similar to this:

MINITAB®
For Windows

The Student Version of MINITAB for Windows 95 / Windows NT
Based on MINITAB Release 12

This software is licensed for use by:
cbrase
acc
WINSV12.11

Minitab Inc.
http://www.minitab.com

Copyright (C) 1998, Minitab Inc.

After a pause, a split screen appears. The session window is at the top, and the data window is at the bottom.

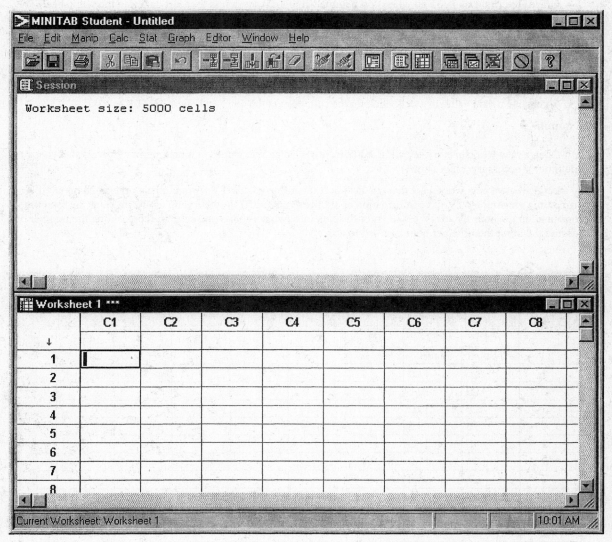

Notice the main menu items:

File   Edit   Manip   Calc   Stat   Graph   Editor   Window   Help

The toolbar contains icons for frequently used operations.

We enter data in the individual cells in the Data window. We can enter commands directly in the Session window, or use the menu options together with dialog boxes to generate the commands.

To end MINITAB: Click on the File option. Select Exit and click on it or press ENTER.

Menu selection summary: ➤**File**➤**Exit**

Entering Data

One of the first tasks you do when you begin a MINITAB session is to insert data into the worksheet. The easiest way to enter data is to enter it directly into the Data worksheet. Notice that the active cell is outlined by a heavier box.

To enter a number, type it in the active box and then press ENTER or TAB. The data value is entered and the next cell is activated. Data for a specific variable are usually entered by column. Notice that the there is a cell for a column label above row number 1.

To change a data value in a cell, click on the cell, correct the data, and press ENTER or TAB.

Example

Open a new worksheet by selecting ➤**File**➤**New.** (Note: You can have a maximum of 5 worksheets open at one time. If necessary, close some.)

Let's create a new worksheet that has data in it regarding ads on TV. A random sample of 20 hours of prime time viewing on TV gave information about the number of TV ads in each hour as well as the total time consumed in the hour by ads. We will enter the data into two columns: one column representing the number of ads and the other the time per hour devoted to ads.

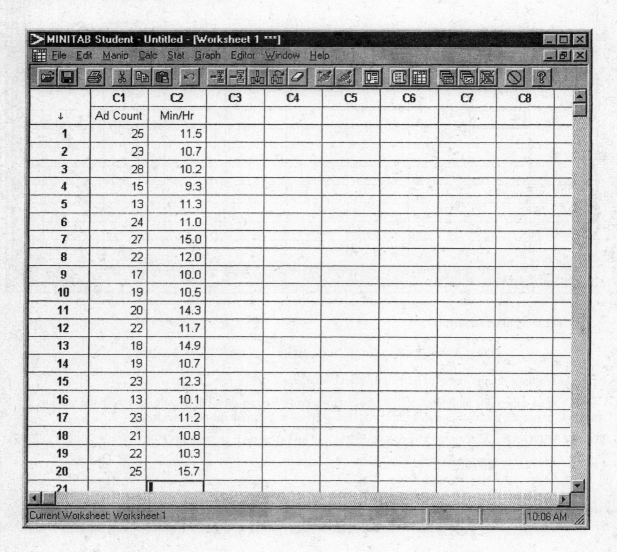

	C1	C2	C3	C4	C5	C6	C7	C8
↓	Ad Count	Min/Hr						
1	25	11.5						
2	23	10.7						
3	28	10.2						
4	15	9.3						
5	13	11.3						
6	24	11.0						
7	27	15.0						
8	22	12.0						
9	17	10.0						
10	19	10.5						
11	20	14.3						
12	22	11.7						
13	18	14.9						
14	19	10.7						
15	23	12.3						
16	13	10.1						
17	23	11.2						
18	21	10.8						
19	22	10.3						
20	25	15.7						
21								

Current Worksheet: Worksheet 1   10:06 AM

Notice that we typed a name for each column. Try to keep the name at 8 characters or fewer. Longer names may be truncated in data displays.

To switch between the Data window and the Session window, click on Window and select the window you want. The Data window is called the Worksheet window.

Working with Data

There are several commands for inserting or deleting rows or cells. One way to access these commands is to use the Manip menu option or the Edit menu option.

Click on the Manip menu item or the Edit menu item. You will see these cascading options.

Manip	Edit
Subset Worksheet	Clear Cells
Split Worksheet	Delete Cells
Sort	Copy Cells
Rank	Cut Cells
Delete rows	Select All Cells
Erase variables	Edit Last Dialog
Copy Columns	Command Line Editor
Stack/Unstack	Preferences
Concatenate	
Code	
Change Data Type	
Display Data	

A useful item is **Change Data Type.** If you accidentally typed a letter instead of a number, you have changed the data type to alpha. To change it back to numeric, use ➤**Manip**➤**Change Data Type** and fill in the dialog box. The same process can be used to change back to alpha.

If you want to see the data displayed in the session window, select ➤**Manip**➤**Display Data** and select the columns you want to see displayed.

To print the worksheet, click the cell in the upper-left corner of the Data window. Press [Shift] + [Ctrl] + [End] to highlight the entire Data window containing all your data. Then click on the printer icon on the toolbar. You can also select ➤**File**➤**Print Worksheet** from the menus.

Manipulating Data

You can also do calculations with entire columns. Click on the $\underline{C}$alc menu item and select Calculator (➤**Calc**➤**Calculator**). The dialog box appears:

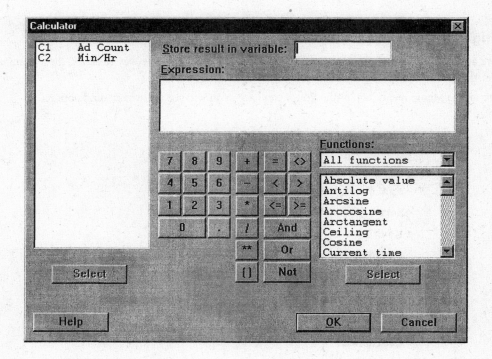

You can store the results in a new column, say C3. To multiply each entry from C1 by 3 and add 4, type 3, click on the multiply key * on the calculator, type C1, click on the + key on the calculator, type 4. Click on OK. The results of this arithmetic will appear in column C3 of the data sheet.

Saving a Worksheet

Click on the $\underline{F}$ile menu and select Save Current Worksheet As. A dialog box similar to the following appears. (Select ➤**File**➤**Save Current Worksheet As.**)

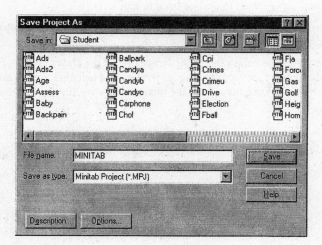

For most computer labs, you will save your file on a 3½ inch floppy. Insert it in the appropriate drive. Scroll down the Save in button until you find 3½ Floppy (A:). Then select a file name. In most cases you will save the file as a MINITAB file. If you change versions of MINITAB or systems, you might select MINITAB portable.

Example

Let's save the worksheet created in the previous example (information about ads on TV).

If you added Column C3 as described under Manipulating the Data, highlight all the entries of the column and press the Del key. Your worksheet should have only two columns. Use ➤**File**➤**Save Current Worksheets.** Insert a diskette in drive A. Scroll down the Save in box and select 3½ Floppy (A:). Name the file Ads. Click on Save. The worksheet will be saved as Ads.mtw.

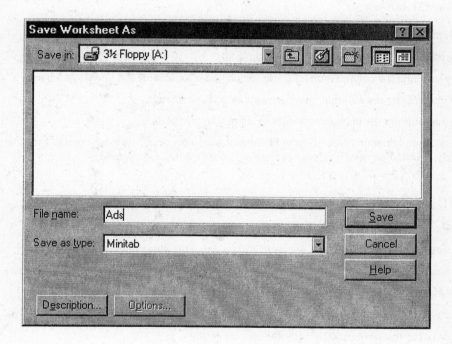

LAB ACTIVITIES FOR GETTING STARTED WITH MINITAB

1. Go to your computer lab (or use your own computer) and learn how to access MINITAB.

2. (a) Use the data worksheet to enter the data:

Enter the data

| | 1 | 3.5 | 4 | 10 | 20 | in Column C1. |

| | 3 | 7 | 9 | 8 | 12 | in Column C2. |

(b) Use ➤**Calc**➤**Calculator** to create C3. The data in C3 should be 2*C1 + C2. Check to see that the first entry in C3 is 5. Do the other entries check?

(c) Name C1 First, C2 Second, C3 Result.

(d) Save the worksheet as Prob 2 on a floppy diskette.

(e) Retrieve the worksheet by selecting ➤**File**➤**Open Worksheet.**

(f) Print the worksheet. Use either the Print Worksheet button or select ➤**File**➤**Print Worksheet.**

RANDOM SAMPLES
(SECTION 1.2 OF *UNDERSTANDING BASIC STATISTICS*)

In MINITAB you can take random samples from a variety of distributions. We begin with one for the simplest: random samples from a range of consecutive integers under the assumption that each of the integers is equally likely to occur. The basic command RANDOM draws the random sample, and subcommands refer to the distribution being sampled. To sample from a range of equally likely integers, we use subcommand INTEGER. The menu selection options are:

➤**Calc**➤**Random Data**➤**Integer**

Dialog Box Responses

Generate _____ rows of data: Enter the sample size.

Store in: Enter the column number C# in which you wish to store the sample numbers.

Minimum: Enter the minimum integer value of your population.

Maximum: Enter the maximum integer value of your population.

The random sample numbers are given in the order of occurrence. If you want them in ascending order (so you can quickly check to see if any values are repeated), use the SORT command.

➤**Manip**➤**Sort**

Dialog Box Responses

Sort columns: Enter the column number C# containing the data you wish to sort.

Store sorted column in: Enter the column number C# where you want to store the sorted data.

This may be the same column number as that containing the original unsorted data.

Sort by column: Enter the same column number C# that contains the original data.

Leave the rest of the sort by columns options empty.

Example

There are 175 students enrolled in a large section of introductory statistics. Draw a random sample of 15 of the students.

We number the students from 1 to 175, so we will be sampling from the integers 1 to 175. We don't want any student repeated, so if our initial sample has repeated values, we will continue to sample until we have 15 distinct students. We sort the data so that we can quickly see if any values are repeated.

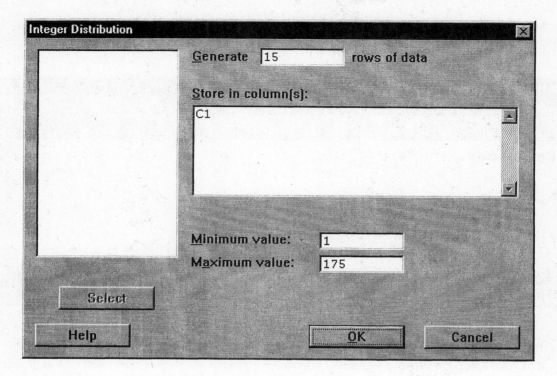

Next, sort the data.

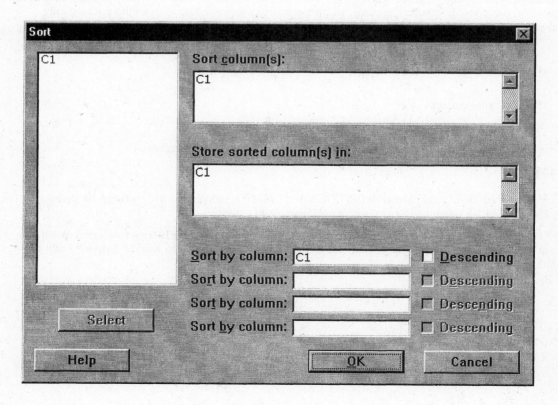

Switch to the Data window and type the name Sample as the header to C1. To display the data, use the command **>Manip>Display Data.** The results are shown. (Your results will vary.)

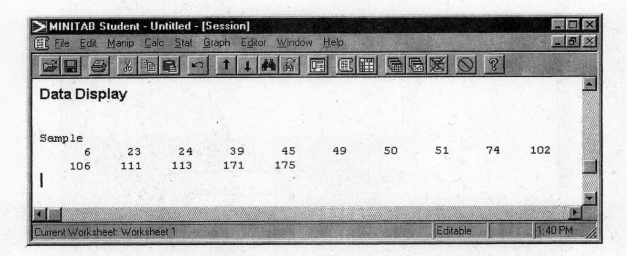

```
>MINITAB Student - Untitled - [Session]

File  Edit  Manip  Calc  Stat  Graph  Editor  Window  Help


Data Display


Sample
       6      23      24      39      45      49      50      51      74     102
     106     111     113     171     175

Current Worksheet: Worksheet 1                              Editable              1:40 PM
```

We see that no data are repeated. If you have repetitions, keep sampling until you get 15 distinct values.

Random numbers are also used to simulate activities or outcomes of a random experiment such as tossing a die. Since the six outcomes 1 through 6 are equally likely, we can use the RANDOM command with the INTEGER subcommand to simulate tossing a die any number of times. When outcomes are allowed to occur repeatedly, it is convenient to tally, count, and give percents of the outcomes. We do this with the tally command and appropriate subcommands.

>Stat>Table>Tally

Dialog Box Responses

Variables: Column number C# or column name containing data

Option to check: counts, percents, cumulative counts, cumulative percents.

Example

Use the RANDOM command with INTEGER A = 1 to B = 6 subcommand to simulate 100 tosses of a fair die. Use the TALLY command to give a count and percent of outcomes.

Generate the random sample using the menu selection **>Random Data>Integer**, with generate at 100, min at 1 and max at 6. Type the name Outcome as the header for C1. Then use **>Stat>Tables>Tally** with counts and percents checked.

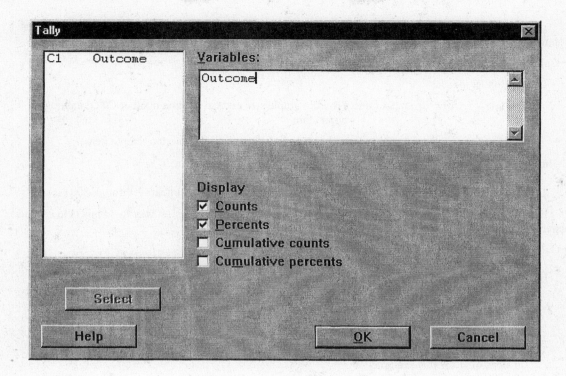

The results are shown. (Your results will vary.)

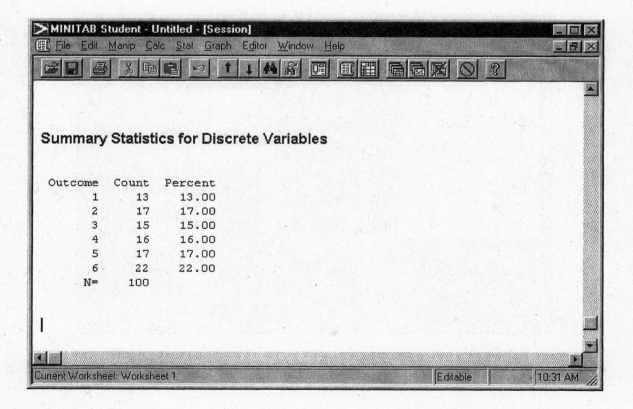

If you have a finite population, and wish to sample from it, you may use the command SAMPLE. This command requires that your population already be stored in a column.

➤**Calc**➤**Random Data**➤**Sample from Columns**

Dialog Box Responses

Sample ____ rows from columns: Provide sample size and list column number C# containing population

Store sample in: Provide column number C# where you want to store the sample items.

Example

Take a sample of size 10 without replacement from the population of numbers 1 through 200.

First we need to enter the numbers 1 through 200 in column C1. The easiest way to do this is to use the patterned data option.

➤**Calc**➤**Make Patterned Data**➤**Simple Set of Numbers**

Dialog Box Responses

Store patterned data in: List column number

From first number: 1 for this example

To last value: 200 for this example

In steps of: 1 for this example

Tell how many times to list each value or sequence.

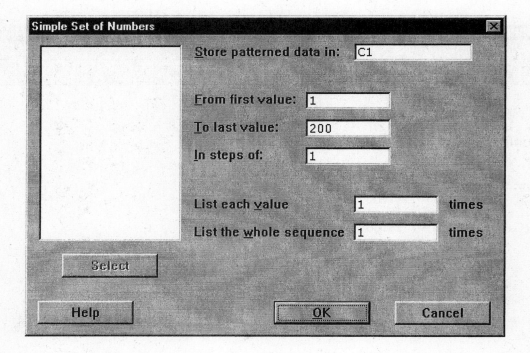

Next we use the **➤Calc➤Random Data➤Sample from Columns** choice to take a sample of 10 items from C1 and store them in C2.

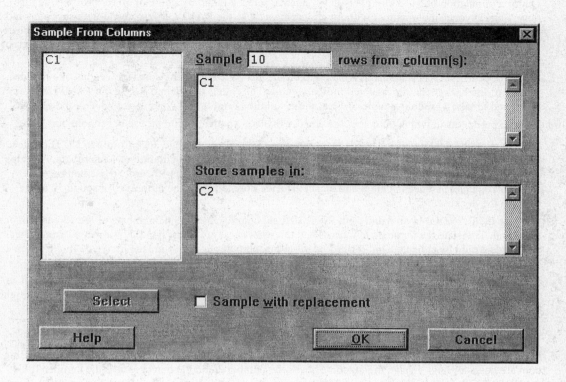

Finally, go to the Data window and label C2 as Sample. Use **➤Minip➤Display Data.** The results are shown. (Your results will vary.)

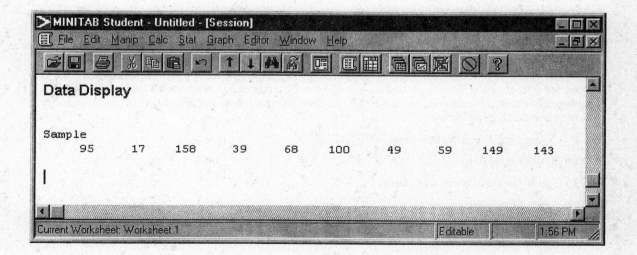

LAB ACTIVITIES FOR RANDOM SAMPLES

1. Out of a population of 8173 eligible count residents, select a random sample of 50 for prospective jury duty. Should you sample with or without replacement? Use the RANDOM command with subcommand INTEGER A = 1 to B = 8173. Use the SORT command to sort the data so that you can check for repeated values. If necessary, use the RANDOM command again to continue sampling until you have 50 different people.

2. Retrieve the MINITAB worksheet Svls02.mtp on the CD-ROM. This file contains weights of a random sample of linebackers on professional football teams. The data is in Column 1. Use the SAMPLE command to take a random sample of 10 of these weights. Print the 10 weights included in the sample.

Simulating experiments in which outcomes are equally likely is another important use of random numbers.

3. We can simulate dealing bridge hands by numbering the cards in a bridge deck from 1 to 52. Then we draw a random sample of 13 numbers without replacement from the population of 52 numbers. A bridge deck has 4 suits: hearts, diamonds, clubs, and spades. Each suit contains 13 cards: those numbered 2 through 10, a jack, a queen, a king, and an ace. Decide how to assign the numbers 1 through 52 to the cards in the deck.

   (a) Use the RANDOM command with INTEGER subcommand to get the numbers of the 13 cards in one hand. Translate the numbers into cards and tell what cards are in the hand. For a second game, the cards would be collected and reshuffled. Use the computer to determine the hand you might get in a second game.

   (b) Store the 52 cards in C1, and then use the SAMPLE command to sample 13 cards. Put the results in C2, name C2 as 'my hand' and print the results. Repeat this process to determine the hand you might get in a second game.

   (c) Compare the four hands you have generated. Are they different? Would you expect this result?

4. We can also simulate the experiment of tossing a fair coin. The possible outcomes resulting from tossing a coin are heads or tails. Assign the outcome heads the number 2 and the outcome tails the number 1. Use RANDOM with INTEGER subcommand to simulate the act of tossing a coin 10 times. Use TALLY with COUNTS and PERCENTS subcommands to tally the results. Repeat the experiment with 10 tosses. Do the percents of outcomes seem to change? Repeat the experiment again with 100 tosses.

COMMAND SUMMARY

Instead of using menu options and dialog boxes, you can type commands directly into the Session window. Notice that you can enter data via the session window with the commands READ and SET rather than through the data window. The following commands will enable you to open worksheets, enter data, manipulate data, save worksheets, etc. Note: Switch to the Session window; the menu choice **➤Editor➤Enable Command Language** allows you to enter commands directly into the Session window and also shows the commands corresponding the to menu choices.

HELP   gives general information about MINITAB

   WINDOWS menu: **Help**

INFO   gives the status of the worksheet

STOP   ends MINITAB session

   WINDOWS menu: **➤File➤Exit**

To Enter Data

READ C...C   puts data into designated columns

READ 'filename' C...C   reads data from file into columns

SET C   puts data into single designated column

SET 'filename' C   reads data from file into column

NAME C = 'name' names column C

   WINDOWS menu: You can enter data in rows or columns and name the column

   in the DATA window. To access the Data window, select **➤Window➤Data**

RETRIEVE 'filename'

   WINDOWS menu: **➤File➤Retrieve**

To Edit Data

LET C(K) = K   changes the value in row K of column C

INSERT K K C C   inserts data between rows K and K into columns C to C

DELETE K K C C   deletes data between rows K and K from column C to C

   WINDOWS menu: You can edit data in rows or columns in the Data window.

             To access the Data window, select **➤Window➤Data.**

COPY C INTO C   copies column C into column C

USE rows K...K   subcommand to copy designated rows

OMIT rows K...K   subcommand to omit designated rows

   WINDOWS menu: **➤Manip➤Copy Columns**

ERASE E...E   erases designated columns or constants

   WINDOWS menu: **➤Manip➤Erase Variables**

To Output Data

PRINT E...E   prints designated columns or constant

   WINDOWS menu: **➤File➤Display Data**

SAVE 'filename' saves current worksheet

PORTABLE   subcommand to make worksheet portable

   WINDOWS menu: **➤File➤Save Worksheet**

   WINDOWS menu: **➤File➤Save Worksheet As...** you may select portable

WRITE 'filename' C...C   saves data in ASCII file

   WINDOWS menu: **➤File➤Other Files➤Export ASCII Data**

Miscellaneous

PAPER   prints session

NOPAPER   stops printing

WINDOWS menu: ➤**File**➤**Print Window**

OUTFILE = 'filename'   saves session in ASCII file

NOOUTFILE   ends OUTFILE

WINDOWS menu: ➤**File**➤**Other Files**➤**Start/Stop Recording**

To Generate a Random Sample

RANDOM K INTO C…C selects a random sample from the distribution described in the subcommand.

 WINDOWS menu: ➤**Cal**➤**Random data**

INTEGER K to K distribution of integers from K to K

Other distributions that may be used with the RANDOM command. We sill study many of these in later chapters.

 BERNOULLI P = K

 BINOMIAL N = K, P = K

 CHISQUARE degrees of freedom = K

 DISCRETE values in C probabilities in C

 F df numerator = K, df denominator = K

 NORMAL [mean = K [standard deviation = k]]

 POISSON mean = K

 T degrees of freedom = K

 UNIFORM continuous distribution on [K to K]

SAMPLE K rows from C…C and put results in C…C takes a random sample of rows without replacement

REPLACE causes the sample to be taken with replacement.

To Organize Data

SORT C…C put in C…C sorts the data in the first column and carries the other columns along.

 WINDOWS menu: ➤**Manip**➤**Sort**

DESCENDING C…C subcommand to sort in descending order

TALLY DATA IN C…C tallies data in columns with integers.

COUNTS

PERCENTS

CUMCOUNTS

ALL gives all four values.

 WINDOWS menu: ➤**Stats**➤**Tables**➤**Tally**

CHAPTER 2 ORGANIZING DATA

HISTOGRAMS
(SECTION 2.2 OF *UNDERSTANDING BASIC STATISTICS*)

MINITAB has graphics in two modes. The default mode is high resolution or professional graphics. There is also an option to use character graphics, which appear in text mode. We will use the high resolution mode.

➤**Graph➤Histogram**

Dialogue Box Responses

    Graph variables: Column containing data

    Data display: Bar for each Graph

    Click on Options and select

        Frequency for type of graph

        Cutpoints for type of interval

        Midpoint/cutpoint positions for Definition of intervals

            List the class boundaries (as computed in *Understandable Statistics*)

Note: If you do not use Options, the computer sets the number of classes automatically. It uses the convention that data falling on a boundary are counted in the class below the boundary.

Example

Let's make a histogram of the data we stored in the worksheet Ads (created in Chapter 1). We'll use C1, the column with the number of ads per hour on prime time TV. Use four classes.

First we need to retrieve the worksheet. Use ➤**File➤Open Worksheet.** Scroll to the drive containing the worksheet. We used 3½ disk drive A. Click on the file.

The number of ads per hour of TV is in column C1. Use ➤**Graph➤Histogram.** The dialogue boxes follow.

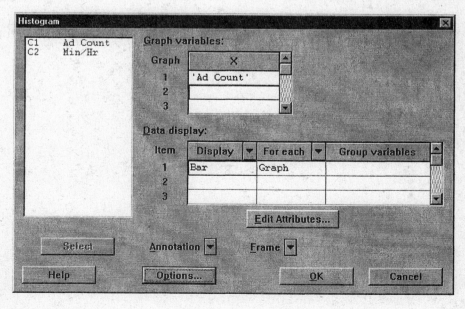

Note that the low data value is 13 and the high value is 28. Using techniques shown in the text *Understanding Basic Statistics*, we see that the class boundaries for 4 classes are 12.5; 16.5; 20.5; 24.5; 28.5. Click on Options and enter these values in the Definition of Intervals. Use spaces to separate the entries. Under Types of Intervals, choose Cutpoint.

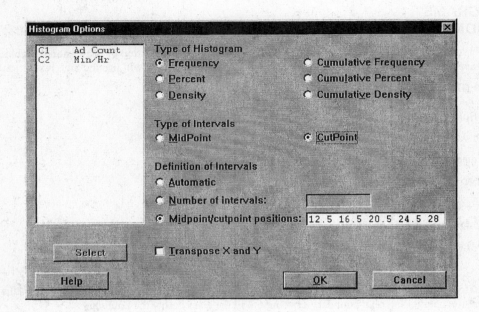

The result follows.

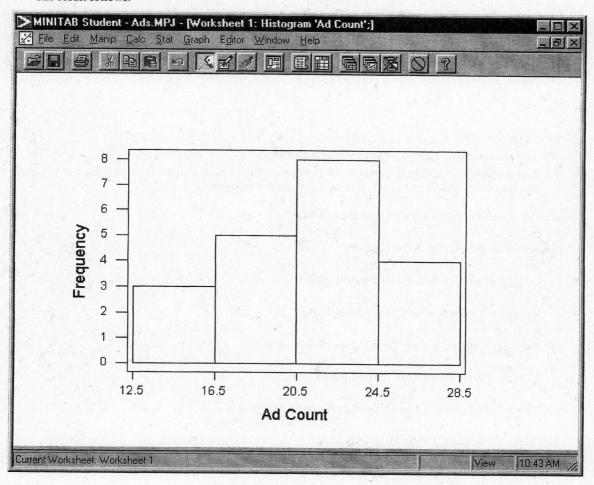

LAB ACTIVITIES FOR HISTOGRAMS

1. The Ads worksheet contains a second column of data that records the number of minutes per hour consumed by ads during prime time TV. Retrieve the Ads worksheet again and use Column C2 to

   (a) make a histogram, letting the computer scale it.

   (b) sort the data and find the smallest data value.

   (c) make a histogram using the smallest data value as the starting value and an increment of 4 minutes. Do this by using cutpoints, with the smallest value as the first cutpoint and incrementing cutpoints by 4 units.

2. As a project for her finance class, Linda gathered data about the number of cash requests made at an automatic teller machine located in the student center between the hours of 6 P.M. and 11 P.M. She recorded the data every day for four weeks. The data values follow.

25	17	33	47	22	32	18	21	12	26	43	25
19	27	26	29	39	12	19	27	10	15	21	20
34	24	17	18								

(a) Enter the data.

(b) Use the command HISTOGRAM to make a histogram.

(c) Use the SORT command to order the data and identify the low and high values. Use the low value as the start value and an increment of 10 to make another histogram.

3. Choose one of the following files from the CD-ROM.

Disney Stock Volume: **Svls01.mtp**

Weights of Pro Football Players: **Svls02.mtp**

Heights of Pro Basketball Players: **Svls03.mtp**

Miles per Gallon Gasoline Consumption: **Svls04.mtp**

Fasting Glucose Blood Tests: **Svls05.mtp**

Number of Children in Rural Canadian Families: **Svls06.mtp**

(a) Make a histogram, letting MINITAB scale it.

(c) Make a histogram using five classes.

4. Histograms are not effective displays for some data. Consider the data

1	2	3	6	7	4	7	9	8	4	12	10

1	9	1	12	12	11	13	4	6	206

Enter the data and make a histogram, letting MINITAB do the scaling. Next scale the histogram with starting value 1 and increment 20. Where do most of the data values fall? Now drop the high value 206 from the data. Do you get more refined information from the histogram by eliminating the high and unusual data value?

STEM-AND-LEAF DISPLAYS
(SECTION 2.3 OF *UNDERSTANDING BASIC STATISTICS*)

MINITAB supports many of the exploratory data analysis methods. You can create a stem-and-leaf display with the following menu choices.

➤**Graph**➤**Stem-and-Leaf**

Dialogue Box Responses

Variables: Column numbers C# containing the data

Increment: Difference in value between smallest possible data in any adjacent lines.

Choose increment 10 for 1 line per stem, or 5 for 2 lines per stem.

Example

Let's take the data in the worksheet Ads and make a stem-and-leaf display of C1. Recall that C1 contains the number of ads occurring in an hour of prime time TV.

Use the menu ➤**Graph**➤**Stem-and-Leaf.**

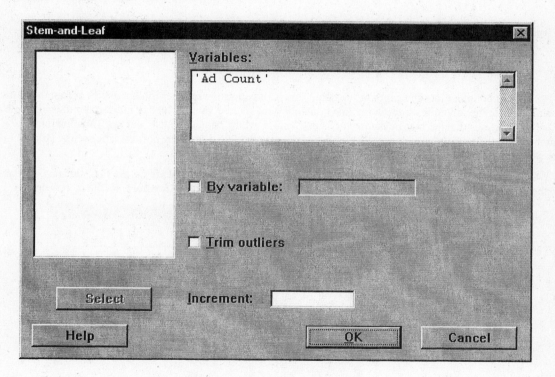

The increment defaulted to 2, so leaf units 0 and 1 are on one line, 2 and 3 on the next, 4 and 5 on the next, etc.

The results follow.

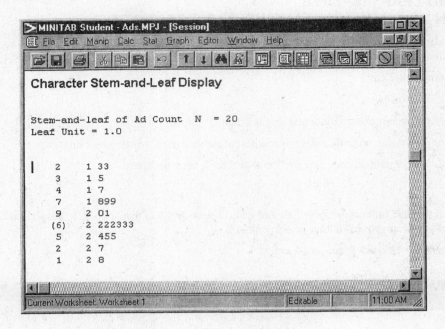

The first column gives the depth of the data. That is, it counts the number of data values accumulated in each line from the top until we reach the line containing the middle value, which is indicated in this example as (6). This indicates that the middle data value is in this line, which contains 6 data values. The remaining numbers in the first column indicate the number of data accumulated from the bottom. The second column gives the stem and the last gives the leaves.

Let's remake a stem leaf with 2 lines per stem. That means that leaves 0–4 are on one line and leaves 5–9 are on the next. The difference in smallest possible leaves per adjacent lines is 5. Therefore, set the increment as 5.

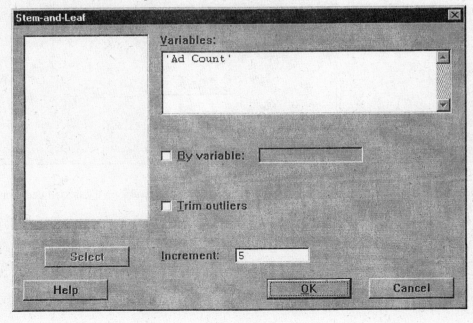

The results follow.

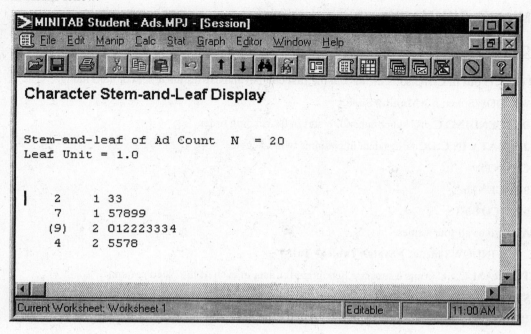

```
MINITAB Student - Ads.MPJ - [Session]

File  Edit  Manip  Calc  Stat  Graph  Editor  Window  Help

Character Stem-and-Leaf Display


Stem-and-leaf of Ad Count   N  = 20
Leaf Unit = 1.0


    2       1 33
    7       1 57899
   (9)      2 012223334
    4       2 5578
```

Current Worksheet: Worksheet 1   Editable   11:00 AM

LAB ACTIVITIES FOR STEM-AND-LEAF DISPLAYS

1. Retrieve worksheet Ads again, and make a stem-and-leaf display of the data in C2. This data gives the number of minutes of ads per hour during prime time TV programs.

   (a) Use an increment of 2.

   (b) Use an increment of 5.

2. In a physical fitness class students ran 1 mile on the first day of class. These are their times in minutes.

12	11	14	8	8	15	12	13	12
10	8	9	11	14	7	14	12	9
13	10	9	12	12	13	10	10	9
12	11	13	10	10	9	8	15	17

   (a) Enter the data in a worksheet.

   (b) Make a stem-and-leaf display and let the computer set the increment.

   (c) Use the TRIM option and let the computer set the increment. How does this display differ from the one in part (b)?

   (d) Set your own increment and make a stem-and-leaf display.

COMMAND SUMMARY

To Organize Data

SORT C…C put in C…C sorts the data in the first column and carries the other columns along.

    WINDOWS menu: ➤**Manip**➤**Sort**

DESCENDING C…C subcommand to sort in descending order

TALLY DATA IN C…C tallies data in columns with integers.

    COUNTS

    PERCENTS

    CUMCOUNTS

    ALL gives all four values.

        WINDOWS menu: ➤**Stats**➤**Tables**➤**Tally**

HISTOGRAM C…C prints a separate histogram for data in each of the listed columns.

    START WITH MIDPOINT = K [end with midpoint = K]

    INCREMENT = K specifies distance between midpoints.

        WINDOWS menu: (for professional graphics) ➤**Graph**➤**Histogram (options for cutpoints)**

        WINDOWS menu: (for character graphics) ➤**Graphs**➤**Character Graphs**➤**Histogram**

STEM-AND-LEAF display of C…C makes separate stem-and-leaf displays of data in each of the listed columns.

    INCREMENT = K sets the distance between two display lines.

    TRIM Outliers lists extreme data on special lines.

        WINDOWS menu: ➤**Graph**➤**Character Graphs**➤**Stem-and-Leaf**

CHAPTER 3 AVERAGES AND VARIATION

AVERAGES AND STANDARD DEVIATION OF UNGROUPED DATA
(SECTIONS 3.1 AND 3.2 OF *UNDERSTANDING BASIC STATISTICS*)

The command DESCRIBE of MINITAB gives many of the summary statistics described in *Understanding Basic Statistics*.

➤**Stat**➤**Basic Statistics**➤**Display Descriptive Statistics** prints descriptive statistics for each column of data.

Dialogue Box Response

Variables: List the columns C1…CN that contain the data.

Graphs option: You may print histograms, etc. directly from this menu.

The labels for Display Descriptive Statistics are

N	number of data in C
N*	number of missing data in C
MEAN	arithmetic mean of C
MEDIAN	median or center of the data in C
TRMEAN	5% trimmed mean (the mean obtained after removing the bottom 5% and top 5% of the data)
STDEV	the sample standard deviation of C, s
SEMEAN	standard error of the mean, **STDEV/SQRT(N)** (we will use this value in Chapter 7)
MIN	minimum data value in C
MAX	maximum data value in C
Q1	1st quartile of distribution in C
Q3	3rd quartile of distribution in C

(**Q1** and **Q3** are similar to Q_1 and Q_3 as discussed in Section 3.4 of *Understandable Statistics*. However, the computation process is slightly different and gives values slightly different from those in the text.)

Example

Let's again consider the data about the number and duration of ads during prime time TV. We will retrieve worksheet Ads and use DESCRIBE on C2, the number of minutes per hour of ads during prime time TV.

First use ➤**File**➤**Open Worksheet** to open worksheet Ads.

Next use ➤**Stat**➤**Basic Statistics**➤**Display Descriptive Statics.** Select **Min/Hr** and click on **OK.**

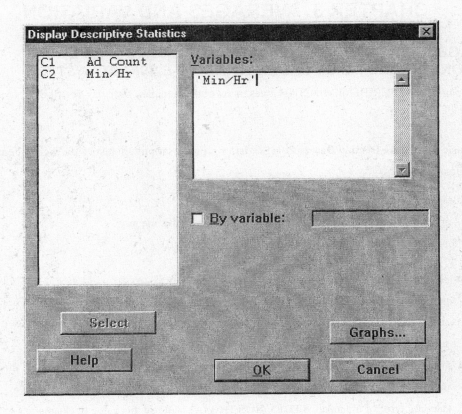

The results follow.

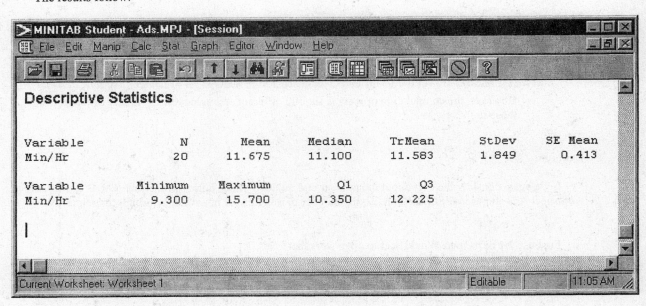

ARITHMETIC IN MINITAB

The standard deviation given in STDEV is the sample standard deviation

$$s = \sqrt{\frac{\sum(x-\bar{x})^2}{N-1}}$$

We can compute the population standard deviation σ by multiplying s by the factor below:

$$\sigma = s\sqrt{\frac{N-1}{N}}$$

MINITAB allows us to do such arithmetic. Use the built-in calculator under menu selection
➤**Calc**➤**Calculator.** Note that * means multiply and ** means exponent.

Example

Let's use the arithmetic operations to evaluate the population standard deviation and population variance
for the minutes per hour of TV ads. Notice that the sample standard deviation $s = 1.849$ and the sample size is
20.

Use the CALCULATOR as follows: Select ➤**Calc**➤**Calculator.** Then enter the expression for the
population variance on the calculator.

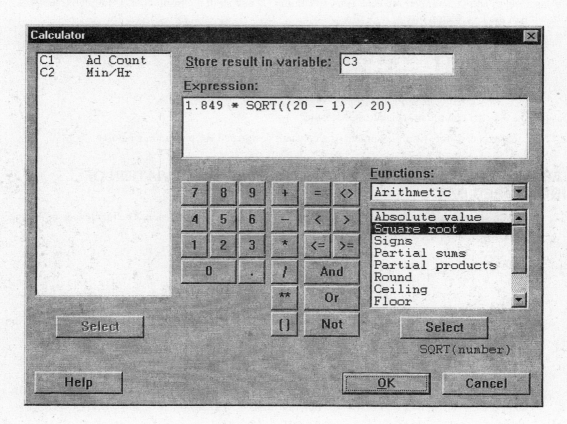

Note that we access the square root function in the function menu. The result is stored in C3.

Go to the Data Window and name C3 as PopStDv. Then use ➤**Manip**➤**Display Data.** The result follows.

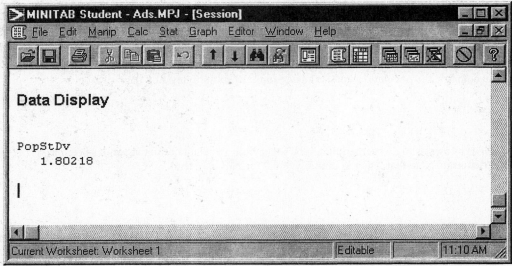

As shown in Chapter 1, you can also use the MINITAB calculator to do arithmetic with columns of data. For instance, to add the number 5 to each entry in column C2 and store the results in C3, simply do the following in the Expression box:

> Type C2 click on + key type
> 5

Designate C3 as the column in which to store results.

Note that you can store a single number as a constant designated K# instead in a column.

LAB ACTIVITIES FOR AVERAGES AND STANDARD DEVIATION OF UNGROUPED DATA

1. A random sample of 20 people were asked to dial 30 telephone numbers each. The incidences of numbers misdialed by these people follow:

3	2	0	0	1	5	7	8	2	6
0	1	2	7	2	5	1	4	5	3

    Enter the data and use the menu selections ➤**Basic Statistics**➤**Display Descriptive Statistics** to find the mean, median, minimum value, maximum value, and standard deviation.

2. Consider the test scores of 30 students in a political science class.

85	73	43	86	73	59	73	84	100	62
75	87	70	84	97	62	76	89	90	83
70	65	77	90	94	80	68	91	67	79

(a) Use the menu selections ➤**Basic Statistics**➤**Display Descriptive Statistics** to find the mean, median, minimum value, maximum value, and standard deviation.

(b) Greg was in political science class. Suppose he missed a number of classes because of illness, but took the exam anyway and made a score of 30 instead of 85 as listed in the data set. Change the 85 (first entry in the data set) to 30 and use the DESCRIBE command again. Compare the new mean, median and standard deviation with the ones in part (a). Which average was most affected: median or mean? What about the standard deviation?

3. Consider the 10 data values

    4   7   3   15   9   12   10   2   9   10

(a) Use the menu selections ➤**Basic Statistics**➤**Display Descriptive Statistics** to find the sample standard deviation of these data values. Then, following example 2 as a model, find the population standard deviation of these data. Compare the two values.

(b) Now consider these 50 data values in the same range.

    7   9   10   6   11   15   17   9   8   2
    2   8   11   15   14   12   13   7   6   9
    3   9   8   17   8   12   14   4   3   9
    2   15   7   8   7   13   15   2   5   6
    2   14   9   7   3   15   12   10   9   10

Again use the menu selections ➤**Basic Statistics**➤**Display Descriptive Statistics** to find the sample standard deviation of these data values. Then, as above, find the population standard deviation of these data. Compare the two values.

(c) Compare the results of parts (a) and (b). As the sample size increases, does it appear that the difference between the population and sample standard deviations decreases? Why would you expect this result from the formulas?

4. In this problem we will explore the effects of changing data values by multiplying each data value by a constant, or by adding the same constant to each data value.

(a) Make sure you have a new worksheet. Then enter the following data into C1:

    1   8   3   5   7   2   10   9   4   6   32

Use the menu selections ➤**Basic Statistics**➤**Display Descriptive Statistics** to find the mean, median, minimum and maximum values and sample standard deviation.

(b) Now use the calculator box to create a new column of data C2 = 10*C1. Use menu selections again to find the mean, median, minimum and maximum values, and sample standard deviation of C2. Compare these results to those of C1. How do the means compare? How do the medians compare? How do the standard deviations compare? Referring to the formulas for these measures (see Sections 3.1 and 3.2 of *Understandable Statistics*), can you explain why these statistics behaved the way they did? Will these results generalize to the situation of multiplying each data entry by 12 instead of 10? Confirm your answer by creating a new C3 that has each datum of C1 multiplied by 12. Predict the corresponding statistics that would occur if we multiplied each datum of C1 by 1000. Again, create a new column C4 that does this, and use DESCRIBE to confirm your prediction.

(c) Now suppose we add 30 to each data value in C1. We can do this by using the calculator box to create a new column of data C6 = C1 + 30. Use menu selection ➤**Basic Statistics**➤**Display Descriptive Statistics** on C6 and compare the mean, median, and standard deviation to those shown for C1. Which are the same? Which are different? Of those that are different, did each change by being 30 more than the corresponding value of part (a)? Again look at the formula for the standard deviation. Can you predict the observed behavior from the formulas? Can you generalize these results? What if we added 50 to each datum of C1? Predict the values for the mean, median, and sample standard deviation. Confirm your predictions by creating a column C7 in which each datum is 50 more than that in the respective position of C1. Use menu selections ➤**Basic Statistic**➤**Display Descriptive Statistics** on C7.

(d) Name C1 as 'orig', C2 as 'T10', C3 as 'T12', C4 as 'T1000', C6 as 'P30', C7 as 'P50'. Now use the menu selections ➤**Basic Statistic**➤**Display Descriptive Statistics** C1-C4 C6 C7 and look at the display. Is it easier to compare the results this way?

BOX-AND-WHISKER PLOTS
(SECTION 3.4 OF *UNDERSTANDING BASIC STATISTICS*)

The box-and-whisker plot is another of the explanatory data analysis techniques supported by MINITAB.

MINITAB uses hinges rather than quartiles Q_1 and Q_3. The lower hinge HL is close in value to Q_1, and the upper hinge HU is close in value to Q_3.

To identify unusual observations, fences located as follows are used:

inner fence: HL – 1.5(HU – HL) and HU + 1.5(HU – HL)

outer fence: HL – 3.0(HU – HL) and HU + 3.0(HU – HL)

Whiskers extend to the most extreme observation within the inner fence. Values between the inner and outer fence are designated by a * while those beyond the outer fences are designated by a 0.

The menu selections are

➤**Graph**➤**Boxplot**

Dialogue Box Responses: Under Y, enter the column number C# containing the data.

Annotation: Open box and you can title the graph.

Display: IQ Range Box with Outliers shown.

There are other boxes available within this box. See Help to learn more about these options.

Example

Now let's make a box-and-whisker plot of the data stored in worksheet ADS. C1 contains the number of ads per hour of prime time TV, while C2 contains the duration per hour of the ads.

Use the menu selection ➤**Graph**➤**Boxplot.** Use C1 for Y. Leave X blank. Click on OK.

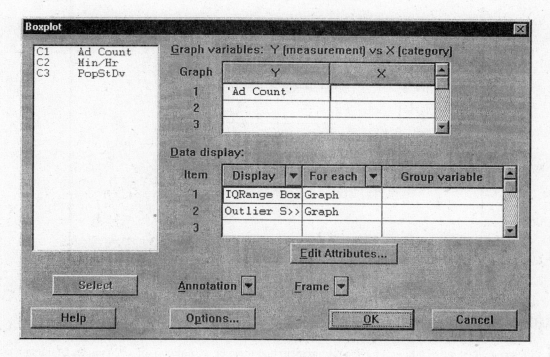

The results follow.

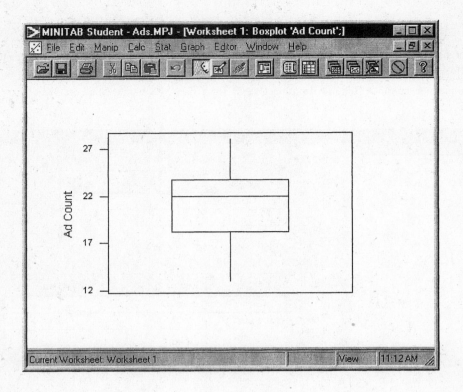

LAB ACTIVITIES FOR BOX-AND-WHISKER PLOTS

1. State-regulated nursing homes have a requirement that there be a minimum of 132 minutes of nursing care per resident per 8-hr shift. During an audit of Easy Life Nursing home, a random sample of 30 shifts showed the number of minutes of nursing care per resident per shift to be

200	150	190	150	175	90	195	115	170	100
140	270	150	195	110	145	80	130	125	115
90	135	140	125	120	130	170	125	135	110

   (a) Enter the data.

   (b) Make a box-and-whisker plot. Are there any unusual observations?

   (c) Make a stem-and-leaf plot. Compare the two ways of presenting the data.

   (d) Make a histogram. Compare the information in the histogram with that in the other two displays.

   (e) Use the ➤**Stat**➤**Basic Statistics**➤**Display Descriptive Statistics** menu selections.

   (f) Now remove any data beyond the outer fences. Do this by inserting an asterisk * in place of the number in the data cell. Use the menu selections ➤**Stat**➤**Basic Statistics**➤**Display Descriptive Statistics** on this data. How do the means compare?

   (h) Pretend you are writing a brief article for a newspaper. Describe the information about the time nurses spend with residents of a nursing home. Use non-technical terms. Be sure to make some comments about the "average" of the data measurements and some comments about the spread of the data.

2. Select one of these data files from the CD-ROM and repeat parts (b) and (h).

> Disney Stock Volume: **Svls01.mtp**
> Weights of Pro Football Players: **Svls02.mtp**
> Heights of Pro Basketball Players: **Svls03.mtp.**
> Miles per Gallon Gasoline Consumption: **Svls04.mtp**
> Fasting Glucose Blood Tests: **Svls05.mtp**
> Number of Children in Rural Canadian Families: **Svls06.mtp**

COMMAND SUMMARY

<u>To Summarize Data by Column</u>

DESCRIBE C...C prints descriptive statistics

      WINDOWS MENU: **➤Stat➤Basic Statistics➤Display Descriptive Statistics**

COUNT	**C [put into K]**	counts the values
N	**C [put into K]**	counts the non-missing values
NMIS	**C [put into K]**	counts the missing values
SUM	**C [put into K]**	sums the values
MEAN	**C [put into K]**	gives arithmetic mean of values
STDEV	**C [put into K]**	gives sample standard deviation
MEDIAN	**C [put into K]**	gives the median of the values
MINIMUM	**C [put into K]**	gives the minimum of the values
MAXIMUM	**C [put into K]**	gives the maximum of the values
SSQ	**C [put into K]**	gives the sum of squares of values

<u>To Summarize Data by Row</u>

RCOUNT	**E...E put into C**
RN	**E...E put into C**
RNMIS	**E...E put into C**
RSUM	**E...E put into C**
RMEAN	**E...E put into C**
RSTDEV	**E...E put into C**
RMEDIAN	**E...E put into C**
RMIN	**E...E put into C**
RMAX	**E...E put into C**
RSSQ	**E...E put into C**

To Display Data

BOXPLOT C   makes a box-and-whisker plot of data in column C

    START = K [end = k]

    INCREMENT = K

    WINDOWS MENU: (professional graphics) **➤Graph➤Boxplot**

To Do Artithmetic

LET E = expression

    evaluates the expression and stored the result in E, where E may be a column or a constant.

    ** raises to a power

    * multiplication

    / division

    + addition

    − subtraction

SQRT(E)   takes the square root

ROUND(E)   rounds numbers to the nearest integer

    Other arithmetic operations are possible.

    WINDOWS menu selections: **➤Calc➤Calculator**

CHAPTER 4 REGRESSION AND CORRELATION

SIMPLE LINEAR REGRESSION: TWO VARIABLES
(SECTIONS 4.1–4.3 OF *UNDERSTANDING BASIC STATISTICS*)

Chapter 4 of *Understanding Basic Statistics* introduces linear regression. Formulas to find the equation of the least-squares line

$$y = a + bx$$

are given in Section 4.2. The formula for the correlation coefficient r and coefficient of determination r^2 are given in Section 4.3. Sections 12.4 and 12.5 continue this discussion with inferences relating to linear regression.

The menu selection ➤**Stat➤Regression➤Regression** gives the equation of the least-squares line, and the value of the coefficient of determination r^2 (R-sq), as well as several other values such as R-sq adjusted (an unbiased estimate of the population r^2). For simple regression with a response variable and one explanatory variable, we can get the value of the Pearson product moment correlation coefficient r by simply taking the square root of R-sq.

The standard deviation, t-ratio and P values of the coefficients are also given. The P value is useful for testing the coefficients to see that the population coefficient is not zero (see Section 12.5 of *Understanding Basic Statistics* for a discussion about testing the coefficients). For the time being we will not use these values.

Depending on the amount of output requested (controlled by the options selected under the **[Results]** button) you will also see an analysis of variance chart, as well as a table of x and y values with the fitted values y_p and residuals $(y - y_r)$. We will not use the analysis of variance chart in our introduction to regression. However, in more advanced treatments of regression, you will find it useful.

To find the equation of the least-squares line and the value of the correlation coefficient, use the menu options

➤**Stat➤Regression➤Regression**

Dialog Box Responses

> Response: Enter the column number C# of the column containing the responses (that is Y values).

> Predictor: Enter the column number C# of the column containing the explanatory variables (that is, X values).

> [Graphs]: Do not click on at this time.

> [Results]: Click on and select the second option, Regression equation, etc.

> [Options]: We will click on this option when we wish to do predictions for new variables.

> [Storage]: Click on and select the fits and residual options if you wish. Residuals are discussed in Section 12.4

To graph the scatter plot and show the least-squares line on the graph, use the menu options

➤**Stat➤Regression➤Regression**

Dialog Box Responses

> Response: List the column number C# of the column containing the Y values.

> Predictor: List the column number C# of the column containing the X values.

> Type of Regression model: Select Linear.

[Options]: Click on and select Display Prediction Band for a specified confidence level of

prediction band. Do not use if you do not want the prediction band.

[Storage]: This button gives you the same storage options as found under regression.

To find the value of the correlation coefficient directly and to find its corresponding *P* value (Section 12.5), use the menu selection

> **➤Stat➤Regression➤Regression**

Dialog Box Responses

Variables: List the column number C# of the column containing the X variable and the column

number C# of the column containing the Y variable.

Select the *P* value option.

Example

Merchandise loss due to shoplifting, damage, and other causes is called shrinkage. Shrinkage is a major concern to retailers. The managers of H.R. Merchandise think there is a relationship between shrinkage and number of clerks on duty. To explore this relationship, a random sample of 7 weeks was selected. During each week the staffing level of sales clerks was kept constant and the dollar value (in hundreds of dollars) of the shrinkage was recorded.

X	10	12	11	15	9	13	8	
Y	19	15	20	9	25	12	31	(in hundreds)

Store the value of X in C1 and name C1 as X. Sore the values of Y in C2 and name C2 as Y.

Use menu choices to give descriptive statistics regarding the values of X and Y. Use commands to draw an (X, Y) scatter plot and then to find the equation of the regression line. Find the value of the correlation coefficient, and test to see if it is significant.

(a) First we will use **➤Stat➤Basic Statistics➤Display Descriptive Statistics** and each of the columns X, and Y. Note that we select both C1 and C2 in the variables box.

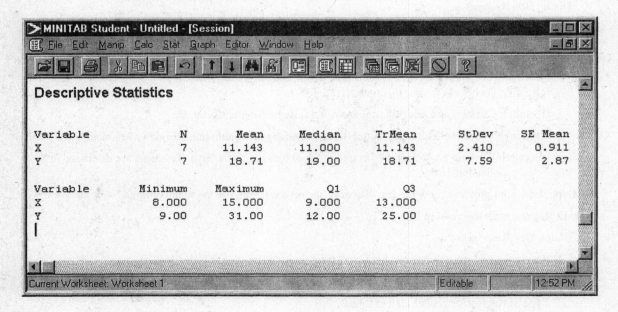

(b) Next we will use ➤**Stat**➤**Regression**➤**Fitted Line Plot** to graph the scatter plot and show the least-squares line on the graph. We will not use prediction bands.

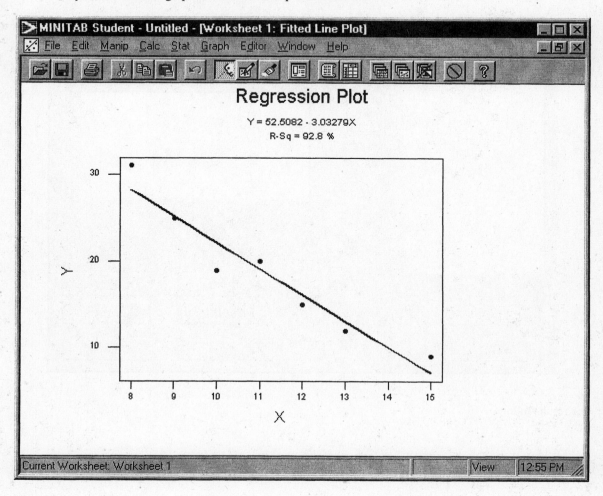

Notice that the equation of the regression line is given on the figure, as well as the value of r^2.

(c) However, to find out more information about the linear regression model, we use the menu selection ➤**Stat**➤**Regression**➤**Regression.** Enter C2 for Response and C1 for Predictor.

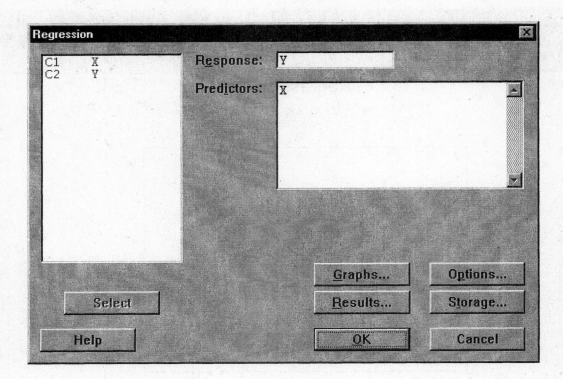

The results follow.

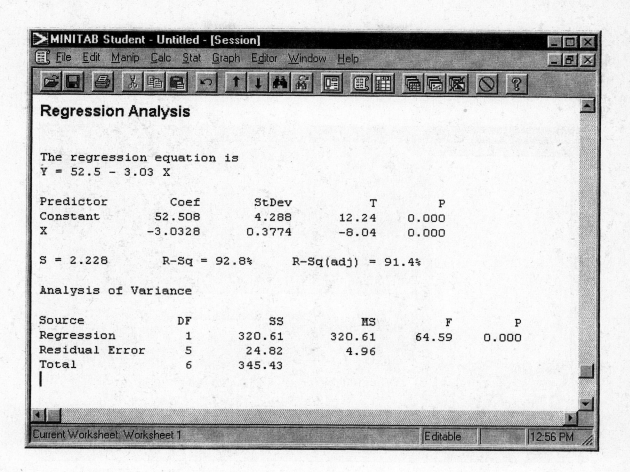

Regression Analysis

The regression equation is
Y = 52.5 - 3.03 X

Predictor	Coef	StDev	T	P
Constant	52.508	4.288	12.24	0.000
X	-3.0328	0.3774	-8.04	0.000

S = 2.228   R-Sq = 92.8%   R-Sq(adj) = 91.4%

Analysis of Variance

Source	DF	SS	MS	F	P
Regression	1	320.61	320.61	64.59	0.000
Residual Error	5	24.82	4.96		
Total	6	345.43			

Notice that the regression equation is given as

$$y = 52.5 - 3.03x$$

We have the value of r^2, R-square = 92.8%. Find the value of r by taking the square root. It is 0.963 or 96.3%.

(d) Next, let's use the prediction option to find the shrinkage when 14 clerks are available.

Use➤**Stat**➤**Regression**➤**Regression.** Your previous selections should still be listed. Now press [Options]. Enter 14 in the prediction window.

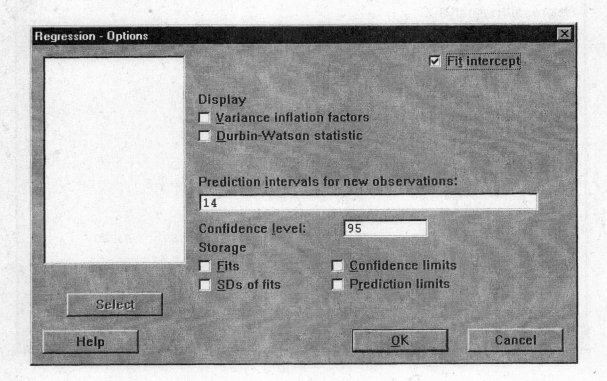

The results follow.

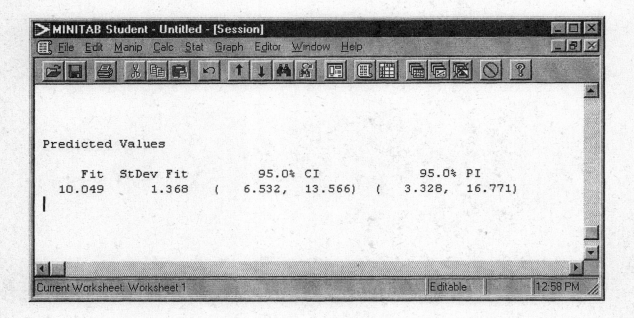

The predicted value of the shrinkage when 14 clerks are on duty is 10.05 hundred dollars, or $1005. A 95% prediction interval goes from 3.33 hundred dollars to 16.77 hundred dollars—that is, from $333 to $1677. (Section 12.4 presents information about confidence intervals for prediction).

(e) Find the correlation coefficient and test it against the hypothesis that there is no correlation. We use the menu options ➤**Stat**➤**Basic Statistics**➤**Correlation.**

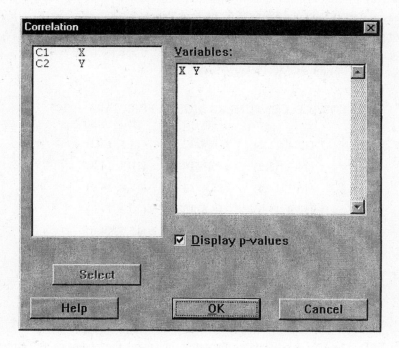

The results are

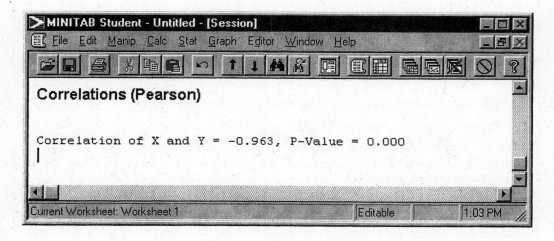

Notice $r = -0.963$. We will study the P value of the correlation in Section 12.5.

LAB ACTIVITIES FOR SIMPLE LINEAR REGRESSION: TWO VARIABLES

1. Open or retrieve the worksheet **Slr01.mtp** from the CD-ROM. This worksheet contains the following data, with the list price in column C1 and the best price in the column C2. The best price is the best price negotiated by a team from the magazine.

   List Price versus Best Price for a New GMC Pickup Truck

   In the following data pairs (X, Y)

   X = List Price (in $1000) for a GMC Pickup Truck

   Y = Best Price (in $1000) for a GMC Pickup Truck

   SOURCE: CONSUMERS DIGEST, FEBRUARY 1994

(12.4, 11.2)	(14.3, 12.5)	(14.5, 12.7)
(14.9, 13.1)	(16.1, 14.1)	(16.9, 14.8)
(16.5, 14.4)	(15.4, 13.4)	(17.0, 14.9)
(17.9, 15.6)	(18.8, 16.4)	(20.3, 17.7)
(22.4, 19.6)	(19.4, 16.9)	(15.5, 14.0)
(16.7, 14.6)	(17.3, 15.1)	(18.4, 16.1)
(19.2, 16.8)	(17.4, 15.2)	(19.5, 17.0)
(19.7, 17.2)	(21.2, (18.6)	

   (a) Use MINITAB to find the least-squares regression line using the best price as the response variable and list price as the explanatory variable.

   (b) Use MINITAB to draw a scatter plot of the data.

   (c) What is the value of r^2, the coefficient of determination? What is the value of r, the correlation coefficient?

   (d) Use the least-squares model to predict the best price for a truck with a list price of $20,000. Note: Enter this value as 20 since X is assumed to be in thousands of dollars.

2. Other MINITAB worksheets appropriate to use for simple linear regression are

   Cricket Chirps Versus Temperature: **Slr02.mtp**

   Source: *The Song of Insects* by Dr. G.W. Pierce, Harvard College Press

   The chirps per second for the striped grouped cricket are stored in C1; the corresponding temperature in degrees Fahrenheit is stored in C2.

   Diameter of Sand Granules Versus Slope on a Beach:

   Slr03.mtp; source *Physical Geography* by A.M. King, Oxford press

   The median diameter (mm) of granules of sand in stored in C1; the corresponding gradient of beach slope in degrees is stored in C2.

   National Unemployment Rate Male Versus Female: **Slr04.mtp**

   Source: *Statistical Abstract of the United States*

   The national unemployment rate for adult males is stored in C1; the corresponding unemployment rate for adult females for the same period of time is stored in C2.

The data in these worksheets are described in the Appendix of this *Guide*. Select these worksheets and repeat parts (a)–(c) of problem 1, using C1 as the explanatory variable and C2 as the response variable.

3. A psychologist interested in job stress is studying the possible correlation between interruptions and job stress. A clerical worker who is expected to type, answer the phone and do reception work has many interruptions. A store manager who has to help out in various departments as customers make demands also has interruptions. An accountant who is given tasks to accomplish each day and who is not expected to interact with other colleagues or customers except during specified meeting times has few interruptions. The psychologist rated a group of jobs for interruption level. The results follow, with X being interruption level of the job on a scale of 1 to 20, with 20 having the most interruptions, and Y the stress level on a scale of 1 to 50, with 50 the most stressed.

Person		1	2	3	4	5	6	7	8	9	10	11	12
	X	9	15	12	18	20	9	5	3	17	12	17	6
	Y	20	37	45	42	35	40	20	10	15	39	32	25

(a) Enter the X values into C1 and the Y values into C2. Use the menu selections ➤**Stat**➤**Basic Statistics**➤**Display Descriptive Statistics** on the two columns. What is the mean of the Y-values? Of the X-values? What are the respective standard deviations?

(b) Make a scatter plot of the data using the ➤**Stat**➤**Regression**➤**Fitted Line** menu selection. From the diagram do you expect a positive or negative correlation?

(c) Use the ➤**Stat**➤**Basic Statistics**➤**Correlation** menu choices to get the value of r. Is this value consistent with your response in part (b)?

(d) Use the ➤**Stat**➤**Regression**➤**Regression** menu choices with Y as the response variable and X as the explanatory variable. Use the [Option] button with predictions 5, 10, 15, 20 to get the predicted stress level of jobs with interruption levels of 5, 10, 15, 20. Is R-sq equal to the square of r as you found in part (c)? What is the equation of the least-squares line?

(e) Redo the ➤**Stat**➤**Regression**➤**Regression** menu option, this time using X as the response variable and Y as the explanatory variable. Is the equation different than that of part (d)? Did R-sq change?

4. The researcher of problem 3 was able to add to her data. Another random sample of 11 people had their jobs rated for interruption level and were then evaluated for stress level.

Person		13	14	15	16	17	18	19	20	21	22	23
	X	4	15	19	13	10	9	3	11	12	15	4
	Y	20	35	42	37	40	23	15	32	28	38	12

Add this data to the data in problem 3, and repeat parts (a) through (e). Does the value of R-sq change? Does the value of r change? Does the least-squares line change?

COMMAND SUMMARY

<u>To Perform Simple Regression</u>

REGRESS C on K explanatory variables in C...C

does regression with the first column containing the response variable, K explanatory variables in the remaining columns.

PREDICT E...E predicts the response variable for the given values of the explanatory variable(s).

RESIDUALS put into C stores the residuals in column C.

WINDOWS menu selection: ➤**Stat**➤**Regression**➤**Regression**

Use the dialog box to list the response and explanatory (prediction) variables. Mark the residuals box. In the Options dialog box list the values of the explanatory variable(s) for which you wish to make a prediction. Select the P.I. confidence interval.

BRIEF K controls the amount of output for K = 1, 2, 3 with 3 giving the most output.

This command is not available from a menu.

There are other subcommands for REGRESS. See the MINITAB Help for your release of MINITAB for a list of the subcommands and their descriptions.

<u>To Find the Pearson Product Moment Correlation Coefficient</u>

CORRELATION for C...C calculates the correlation coefficient for all pairs of columns.

WINDOWS menu selection: ➤**Stat**➤**Basic Statistics**➤**Correlation**

<u>To Graph the Scatter Plot for Simple Regression</u>

With **GSTD** use the **PLOT C vs C** command.

WINDOWS menu selection (MINITAB SE for Windows or MINITAB 10 for Windows):

➤**Stat**➤**Regression**➤**Fitted Line Plot**

CHAPTER 5 ELEMENTARY PROBABILITY THEORY

RANDOM VARIABLES AND PROBABILITY

MINITAB supports random samples from a column of numbers or from many probability distributions. See the options under **>Calcu>Random Data.** By using some of the same techniques shown in Chapter 2 of this guide for random samples, you can simulate a number of probability experiments.

Example

Simulate the experiment of tossing a fair coin 200 times. Look at the percent of heads and the percent of tails. How do these compare with the expected 50% of each?

Assign the outcome heads to digit 1 and tails to digit 2. We will draw a random sample of size 200 from the distributions of integers from a minimum of 1 to a maximum of 2. Use the menu selections **>Calc>Random Data>Integer.** In the dialog box, enter 200 for the number of rows, 1 for the minimum and 2 for the maximum. Put the data in column C1 and label the column Coin. To tally the results use **>Stat>Tables>Tally** and check the counts and percents options. The results are shown. (Your results will vary.)

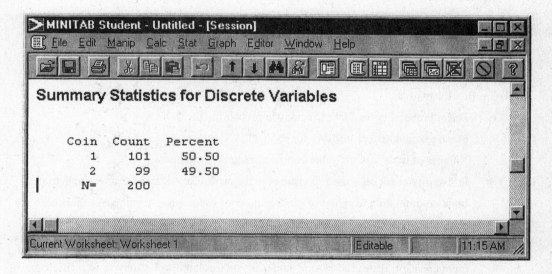

LAB ACTIVITIES FOR RANDOM VARIABLES AND PROBABILITY

1. Use the RANDOM command and INTEGER A = 0 to B = 1 subcommand to simulate 50 tosses of a fair coin. Use the TALLY command with COUNT and PERCENT subcommands to record the percent of each outcome. Compare the result with the theoretical expected percents (50% heads, 50% tails). Repeat the process for 1000 trials. Are these outcomes closer the results predicted by the theory?

2. We can use the RANDOM 50 C1 C2 command with INTEGER A = 1 to B = 6 subcommand to simulate the experiment of rolling two dice 50 times and recording each sum. This command puts outcomes of die 1 into C1 and those of die 2 into C2. Put the sum of the dice into C3. Then use the TALLY command with COUNT and PERCENT subcommands to record the percent of each outcome. Repeat the process for 1000 rolls of the dice.

CHAPTER 6 THE BINOMIAL PROBABILITY DISTRIBUTION AND RELATED TOPICS

BINOMIAL PROBABILITY DISTRIBUTIONS
(SECTIONS 6.2 AND 6.3 OF *UNDERSTANDING BASIC STATISTICS*)

The binomial probability distribution is a discrete probability distribution controlled by the number of trials, n, and the probability of success on a single trial, p.

MINITAB has three main commands for studying probability distributions.

The PDF (probability density function) gives the probability of a specified value for a discrete distribution.

The CDF (cumulative distribution function) for a value X gives the probability a random variable with distribution specified in a subcommand is less than or equal to X.

The INVCDF gives the inverse of the CDF. In other words, for a probability P, INVCDF returns the value X such that $P \approx CDF(X)$. In this case of a binomial distribution. INVCDF often gives the two values of X for which P lies between the respective CDF(X).

The three commands PDF, CDF, and INVCDF apply to many probability distributions. To apply them to a binomial distribution, we need to use the menu selections.

Calc➤Probability distributions➤Binomial

Dialog Box Responses

Select Probability for PDF; Cumulative probability for CDF;

Inverse cumulative probability for INVCDF

Number of trials: use the value of n in a binomial experiment,

Probability of success: use the value of p, the probability of success on a single trial,

Input column: Put the values of r, the number of successes in a binomial experiment in a

column such as C1. Select an optional storage column.

Note: MINITAB uses X instead of r to count the number of successes,

Input constant: Instead of putting values of r in a column, you can type a specific value of

r in the dialog box.

Example

A surgeon performs a difficult spinal column operation. The probability of success of the operation is $p = 0.73$. Ten such operations are scheduled. Find the probability of success for 0 through 10 successes out of these ten operations.

First enter the possible values of r, 0 through 10, in C1 and name the column r. We will put the probabilities in C2, so name the column $P(r)$.

Fill in the dialog box as shown below.

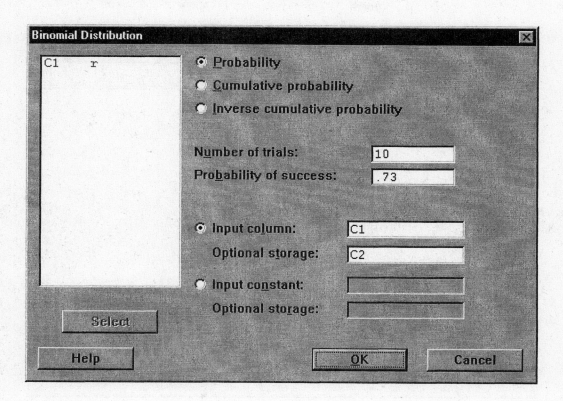

Then use the ➤**Manip**➤**Display data** command.

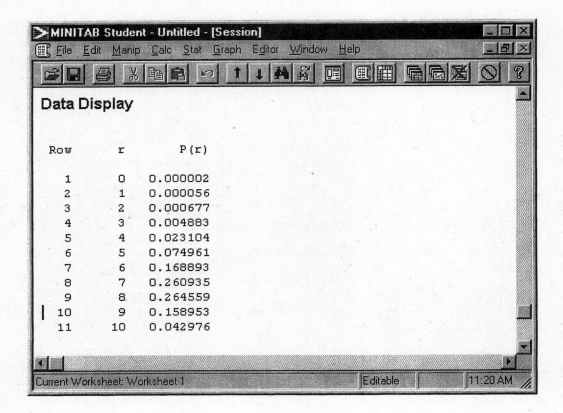

Next use the CDF command to find the probability of 5 or fewer successes. In this case use the option for an input constant of 5. Leave Optional storage blank. The output will be $P(r \le 5)$. Note that MINITAB uses X in place of r.

The results follow.

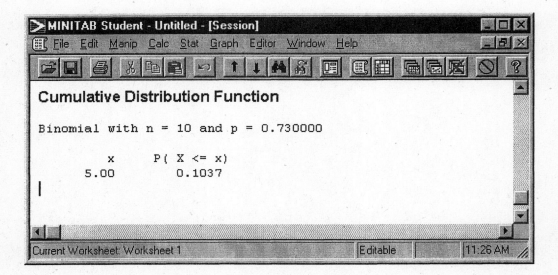

Finally use INVCDF to determine how many operations should be performed in order for the probability of that many or fewer successes to be 0.5. We select Inverse cumulative probability. Use .5 as the input constant.

The results follow.

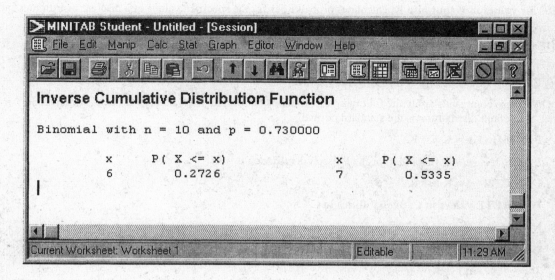

LAB ACTIVITIES FOR BINOMIAL PROBABILITY DISTRIBUTIONS

1. You toss a coin 8 times. Call heads success. If the coin is fair, the probability of success P is 0.5. What is the probability of getting exactly 5 heads out of 8 tosses? of exactly 20 heads out of 100 tosses?

2. A bank examiner's record shows that the probability of an error in a statement for a checking account at Trust Us Bank is 0.03. The bank statements are sent monthly. What is the probability that exactly two of the next 12 monthly statements for our account will be in error? Now use the CDF option to find the probability that <u>at least</u> two of the next 12 statements contain errors. Use this result with subtraction to find the probability that <u>more than</u> two of the next 12 statements contain errors. You can use the Calculator key to do the required subtraction.

3. Some tables for the binomial distribution give values only up to 0.5 for the probability of success p. There is symmetry to the values for p greater than 0.5 with those values of p less than 0.5.

   (a) Consider the binomial distribution with $n = 10$ and $p = .75$. Since there are 0–10 successes possible, put $0 - 10$ in C1. Use PDF option with C1 and store the distribution probabilities in C2. Name C2 = 'P = .75'. We will print the results in part (c).

   (b) Now consider the binomial distribution with $n = 10$ and $p = .25$. Use PDF option with C1 and store the distribution probabilities in C3. Name C3 = 'P = .25'.

   (c) Now display C1 C2 C3 and see if you can discover the symmetries of C2 with C3. How does $P(K = 4$ successes with $p = .75)$ compare to $P(K = 6$ successes with $p = .25)$?

   (d) Now consider a binomial distribution with $n = 20$ and $p = .035$. Use PDF on the number 5 to get $P(K = 5$ successes out of 20 trials with $p = .35)$. Predict how this result will compare to the probability $P(K = 15$ successes out of 20 trials with $p = .65)$. Check your prediction by using the PDF on 15 with the binomial distribution $n = 20$, $p = .65$.

COMMAND SUMMARY

To Find Probabilities

PDF for values in E [put into E] calculates probabilities for the specified values of a discrete distribution and calculates the probability density function for a continuous distribution.

CDF for values in E...E [put into E...E] gives the cumulative distribution. For any value X, CDF X gives the probability that a random variable with the specified distribution has a value less than or equal to X.

INVCDF for values in E [put into E] gives the inverse of the CDF.

Each of these commands apply the following distributions (as well as some others). If no subcommand is used, the default distribution is the standard normal.

BINOMIAL $n = K$ $p = K$

POISSON $\mu = K$ (note that for the Poisson distribution $\mu = \lambda$)

INTEGER $a = K$ $b = K$

DISCRETE values in C, probabilities in C

NORMAL $\mu = K$ $\sigma = K$

UNIFORM $a = k$ $b = K$

T $d.f. = K$

F $d.f.$ **numerator** $= K$ $d.f.$ **denominator** $= K$

CHISQUARE $d.f. = K$

WINDOWS menu selection: ➤**Calc**➤**Probability Distribution**➤**Select distribution**

In the dialog box, select **Probability for PFD; Cumulative probability for CDF; Inverse cumulative for INV;** Enter the required information such as **E, n, p, or μ, d.f.** and so forth.

CHAPTER 7 NORMAL DISTIBUTIONS

GRAPHS OF NORMAL DISTRIBUTIONS
(SECTION 7.1 OF *UNDERSTANDING BASIC STATISTICS*)

The normal distribution is a continuous probability distribution determined by the value of μ and σ. We can sketch the graph of a normal distribution by using the menu selection

➤**Cal**➤**Probability Distribution**➤**Normal**

Dialog Box Responses

Select Probability density for **PDF,** Cumulative probability for **CDF,** or Inverse cumulative probability for INVCDF.

Enter the mean.

Enter the standard deviation.

Select an input column: Put the value of x for which you want to compute P(x) in the designated column. Designate an optional storage column.

Select an input column: If you wish to compute P(x) for a single value x, enter value as the constant.

We will use the normal probability density option **PDF** to create a graph of a normal distribution with a specified mean and standard deviation.

Menu Options for Graphs

To graph functions in MINITAB, we use the menu ➤**Graph**➤**Plot.**

Dialog Box Responses

Specify which column contains data for the X values.

Specify which column contains data for the Y values.

Under Display, scroll to the Connect option.

There are other options available. See the Help menu for more information.

Example

Graph the normal distribution with mean $\mu = 10$ and standard deviation $\sigma = 2$.

Since most of the normal curve occurs over the values $\mu - 3\sigma$ to $\mu + 3\sigma$, we will start the graph at $10 - 3(2) = 4$ and end it at $10 + 3(2) = 16$. We will let MINITAB set the scale on the vertical axis.

To graph a normal distribution, put X values into C1 and Y values (height of the graph at point X) into C2.

Use the menu option ➤**Calc**➤**Make Patterned Data**➤**Simple set of numbers.** Store values in C1. First value is 4, last value is 16, and increment is 0.25. Variables and sequences occur one time each. Name C1 as X.

Use ➤**Calc**➤**Probability Distributions**➤**Normal** to generate the Y values which we will store in C2. Name C2 as P(X).

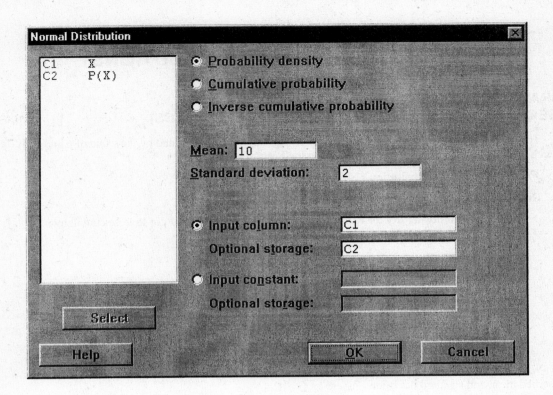

Finally, use **➤Graph➤Plot.** Set the options as shown.

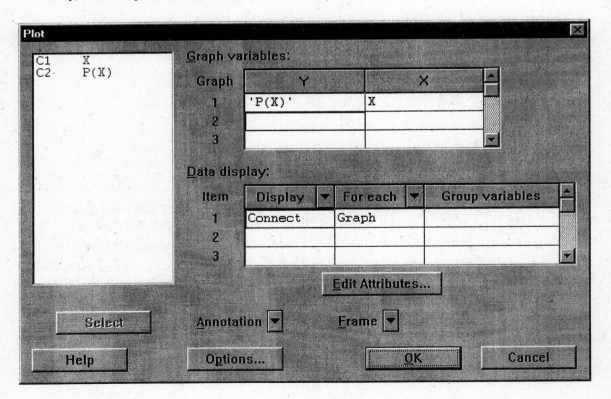

The graph follows.

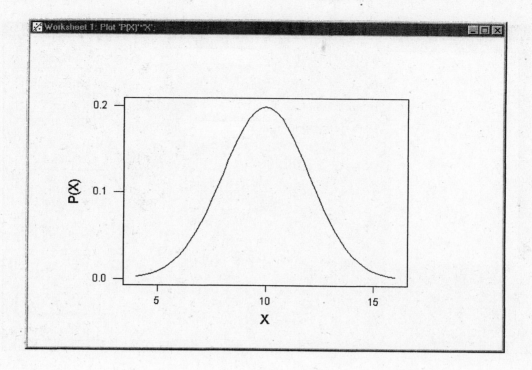

CONTROL CHARTS
(SECTION 6.1 OF *UNDERSTANDABLE STATISTICS*)

MINITAB supports a variety of control charts. The type discussed in Section 6.1 of *Understandable Statistics* is called an individual chart. The menu selection is ➤**Stat**➤**Control Chart**➤**Individual.**

Dialog Box Responses

Variable: Designate the column number C1 where the data is located.

Enter values for the historical mean and standard deviation.

Tests option button lists out-of-control tests. Select numbers 1, 2 and 5 for signals discussed in *Understandable Statistics.*

For information about the other options, see the Help menu.

Example

In a packaging process, the weight of popcorn that is to go in a bag has a normal distribution with $\mu = 20.7$ oz and $\sigma = 0.7$ oz. During one session of packaging, eleven samples were taken. Use an individual control chart to show these observations. The weights were (in oz).

19.5   20.3   20.7   18.9   9.5   20.5
20.7   21.4   21.9   22.7   23.8

Enter the data in column C1, and name the column oz.

Select **Stat➤Control Charts➤Individuals.** Fill in the values for μ and σ.

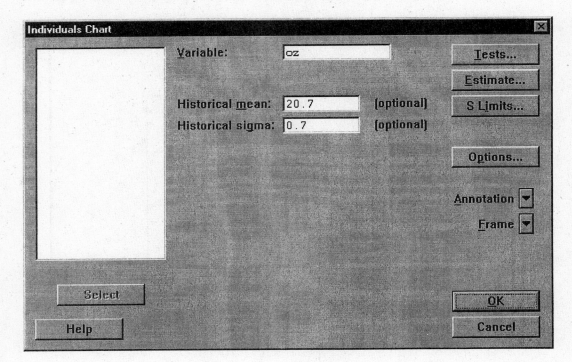

The control chart is

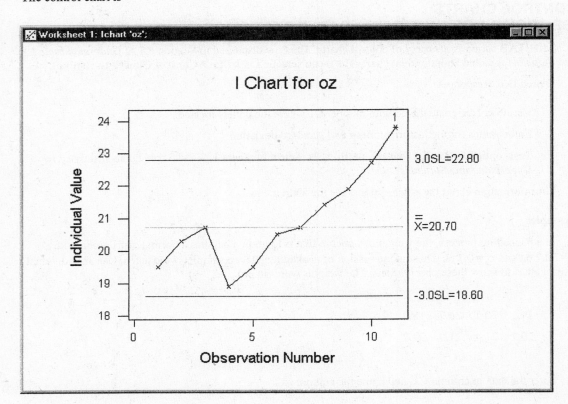

LAB ACTIVITIES FOR GRAPHS OF NORMAL DISTRIBUTIONS AND CONTROL CHARTS

1. (a) Sketch a graph of the standard normal distribution with $\mu = 0$ and $\sigma = 1$. Generate C1 using data from -3 to 3 in increments of 0.5 (generates data from -3 to 3 with .5 increments).

    (b) Sketch a graph of a normal distribution with $\mu = 10$ and $\sigma = 1$. Generate C1 using data from -7 to 7 in increments of 0.5. Compare this graph to that of part (a). Do the height and spread of the graphs appear to be the same? What is different? Why would you expect this difference?

    (c) Sketch a graph of a normal distribution with $\mu = 0$ and $\sigma = 2$. Generate C1 using data from -6 to 6 in increments of 0.5. Compare this graph to that of part (a). Do the height and spread of the graphs appear to be the same? What is different? Why would you expect this difference? Note, to really compare the graphs, it is best to graph them using the same scales. Redo the graph of part (a) using X from -6 to 6. Then redo the graph in this part using the same X values as in part (a) and Y values ranging from 0 to the high value of part (a).

2. Use one of the following MINITAB portable worksheets found on the CD-ROM. In each of the files the target value for the mean μ is stored in the C2(1) position and the target value for the standard deviation is stored in the C3(1) position.

    Yield of Wheat: Tscc01.mtp

    PepsiCo Stock Closing Prices: Tscc02.mtp

    PepsiCo Stock Volume of Sales: Tscc03.mtp

    Futures Quotes for the Price of Coffee Beans: Tscc04.mtp

    Incidence of Melanoma Tumors: Tscc05.mtp

    Percent Change in Consumer Price Index: Tscc06.mtp

    Use the targeted μ and σ values.

COMMAND SUMMARY

Graphing Commands

Character Graphics Commands

PLOT C versus C prints a scatter plot with the first column on the vertical axis and the second on the horizontal axis. The following subcommands can be used with PLOT.

> **TITLE = 'test'** gives a title above the graph.

> **FOOTNOTE = 'test'** places a line of text below the graph.

> **XLABEL = 'test'** labels the x-axis.

> **YLABEL = 'test'** labels the y-axis.

> **SYMBOL = 'symbol'** selects the symbol for the points on the graph. The default is *.

> **XINCREMENT = K** is the distance between tick marks on x-axis.

> **XSTART = K [end = k]** specifies the first tick mark and optionally the last one.

> **YINCREMENT = K** is the distance between tick marks on y-axis.

> **YSTART = K [end = K]** specifies the first tick mark and optionally the last one.

WINDOWS menu selection: ➤**Graph**➤**Character Graphs**➤**Scatter Plot**

Titles, labels, and footnotes are in the **Annotate...**option.

Increment and start are in the **Scale** option.

Professional Graphics

Plot C * C prints a scatter plot with the first column on the vertical axis and the second on the horizontal axis. Note that the columns must be separated by an asterisk *.

Connect connects the points with a line.

Other subcommands may be used to title the graph and set the tick marks on the axes. See your MINITAB software manual for details.

WINDOWS menu selection: ➤**Graph**➤**Plot**

Use the dialog boxes to title the graph, label the axes, set the tick marks, and so forth.

See your MINITAB software manual for details.

Control Charts

CHART C...C produces a control chart under the assumption that the data come from a normal distribution with mean and standard deviation specified by the subcommands.

MU K gives the mean of the normal distribution.

SIGMA K gives the standard deviation.

WINDOWS menu selection: ➤**Stat**➤**Control Chart**➤**Individual**

Enter choices for MU and SIGMA in the dialog box.

CHAPTER 8
INTRODUCTION TO SAMPLING DISTRIBUTIONS

Note: This section uses session window commands instead of menu choices

CENTRAL LIMIT THEOREM
(SECTION 8.1 OF *UNDERSTANDING BASIC STATISTICS*)

The Central Limit Theorem says that if x is a random variable with <u>any</u> distribution having mean μ and standard deviation σ, then the distribution of sample means $\bar{x}$ based on random samples of size n is such that for sufficiently large n:

(a) The mean of the $\bar{x}$ distribution is approximately the same as the mean of the x distribution.

(b) The standard deviation of the x distribution is approximately $\sigma/\sqrt{n}$.

(c) The $\bar{x}$ distribution is approximately a normal distribution.

Furthermore, as the sample size n becomes larger and larger, the approximations mentions in (a), (b) and (c) become better.

We can use MINITAB to demonstrate the Central Limit Theorem. The computer does not prove the theorem. A proof of the Central Limit Theorem requires advanced mathematics and is beyond the scope of an introductory course. However, we can use the computer to gain a better understanding of the theorem.

To demonstrate the Central Limit Theorem, we need a specific x distribution. One of the simplest is the <u>uniform probability distribution</u>.

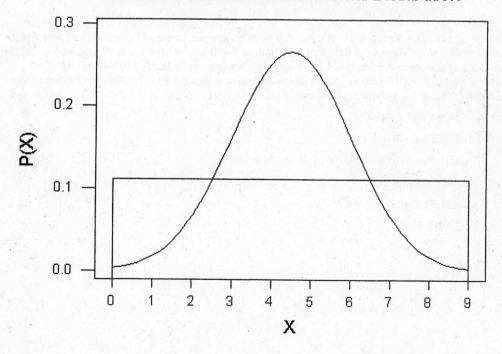

Uniform Distribution and a Normal Distribution

The normal distribution is the usual bell-shaped curve, but the uniform distribution is the rectangular or box-shaped graph. The two distributions are very different.

The uniform distribution has the property that all subintervals of the same length inside the interval 0 to 9 have the same probability of occurrence no matter where they are located. This means that the uniform distribution on the interval from 0 to 9 could be represented on the computer by selecting random numbers from 0 to 9. Since all numbers from 0 to 9 would be equally likely to be chosen, we say we are dealing with a uniform (equally likely) probability distribution. Note that when we say we are selecting random numbers from 0 to 9, we do not just mean whole numbers or integers; we mean real numbers in decimal form such as 2.413912, and so forth.

Because the interval from 0 to 9 is 9 units long and because the total area under the probability graph must by 1 (why?), the height of the uniform probability graph must be 1/9. The mean of the uniform distribution on the interval from 0 to 9 is the balance point. Looking at the Figure, it is fairly clear that the mean is 4.5. Using advanced methods of statistics, it can be shown that for the uniform probability distribution x between 0 and 9,

$$\mu = 4.5 \text{ and } \sigma = 3\sqrt{3}/2 \approx 2.598$$

The figure shows us that the uniform x distribution and the normal distribution are quite different. However, using the computer we will construct one hundred sample means $\bar{x}$ from the x distribution using a sample size of $n = 40$. We can vary the sample size n according to how many columns we use in RANDOM command.

RANDOM 100 C1-C40

UNIFORM a = 0 TO b = 9.

We will see that even though the uniform distribution is very different from the normal distribution, the histogram of the sample means is somewhat bell shaped. Looking at the DESCRIBE command, we will also see that the mean or the $\bar{x}$ distribution is close to the predicted mean of 4.5 and that the standard deviation is close to $\sigma/\sqrt{n}$ or $2.598/\sqrt{40}$ or 0.411.

Example

The following MINITAB program will draw 100 random samples of size 40 from the uniform distribution on the interval from 0 to 9. We put the data into 40 columns. Then we take the mean of each of the 100 rows (40 columns across) and store the result in C50. To do this, we use the RMEAN C1-C40 put into C50 command. Next we DESCRIBE C50 to look at the mean and standard deviation of the distribution of sample means. Finally we look at a histogram of the sample means in C50.

MTB > RANDOM 100 C1-C40;

SUBC > UNIFORM a = 0 to b = 9.

MTB > # Take the mean of the 40 data values in each row.

MTB > # Put the means in C50.

MTB > RMEAN C1-C40 put into C50

MTB > DESCRIBE C50

	N	MEAN	MEDIAN	TRMEAN
C50	100	4.5559	4.5886	4.5531

	STDEV	SEMEAN	MIN	MAX
C50	0.4044	0.0404	3.7003	5.4569

	Q1	Q3
C50	4.1897	4.8487

Note the MEAN and STDEV are very close to the values predicted by the Central Limit Theorem.

MTB > GSTD

MTB > HISTOGRAM C 50

Histogram of C50 N = 100

Midpoint	Count	
3.8	6	******
4.0	10	**********
4.2	12	************
4.4	8	********
4.6	25	*************************
4.8	21	*********************
5.0	11	***********
5.2	2	**
5.4	5	*****

The histogram for this sample does not appear very similar to a normal distribution. Let's try another sample.

	N	MEAN	MEDIAN	TRMEAN
C50	100	4.5265	4.5274	4.5284

	STDEV	SEMEAN	MIN	MAX
C50	0.4021	0.0402	3.4106	5.4145

	Q1	Q3
C50	4.2179	4.7927

MTB > GSTD

MTB > HISTOGRAM C 50

Histogram of C50 N = 100

Midpoint		Count	
3.4	1	*	
3.6	0		
3.8	9	*********	
4.2	14	**************	
4.4	18	******************	
4.6	19	*******************	
4.8	18	******************	
5.0	7	*******	
5.2	6	******	
5.4	3	***	

This histogram looks more like a normal distribution. You will get slightly different results each time you draw 100 samples.

The number of samples used is determined by K, and the size of the samples is determined by the number of columns in the command.

RANDOM K C1-CN

UNIFORM a = 0 TO b = 9.

Be sure that when you take the RMEAN of the rows, you use the same number of columns as you used in the random command

RMEAN C1-CN put into C

Then use the DESCRIBE and HISTOGRAM commands on the column C where you put the means.

You can sample from a variety of distributions, some of which were listed under the RANDOM command in the Chapter 1 Command Summary.

LAB ACTIVITIES FOR CENTRAL LIMIT THEOREM

1. Repeat the experiment of Example 1. That is, draw 100 random samples of size 40 each from the uniform probability distribution between 0 and 9. Then take the means of each of these samples and put the results in C50. Use the commands

> **RANDOM 100 C1-C40**
> **UNIFORM a = 0 to b = 9**
> **RMEAN C1-C40 put into C50**

Next use DESCRIBE on C50. How does the mean and standard deviation of the distribution of sample means compare to those predicted by the Central Limit Theorem? Use HISTOGRAM C50 to draw a histogram of the distribution of sample means. How does it compare to a normal curve? (Note: in the student edition of MINITAB, the worksheet may not be large enough to accommodate this project. Change the sample size to 15 – that is, use C1-C15 in the RANDOM and RMEAN commands.)

2. Next take 100 random samples of size 20 from the uniform probability distribution between 0 and 9. To do this, use only 20 columns in the RANDOM and RMEAN commands. Again put the means in C50, use DESCRIBE and HISTOGRAM on C50 and comment on the results. How do these results compare to those in problem 1? How do the standard deviations compare? (Note: in the student edition of MINITAB, the worksheet may not be large enough to accommodate this project. Change the sample size to 10.)

CHAPTER 9 ESTIMATION

CONFIDENCE INTERVALS FOR A MEAN OR FOR A PROPORTION (SECTIONS 9.1–9.3 OF *UNDERSTANDING BASIC STATISTICS*)

Student's *t* Distribution

In Section 9.1 of *Understanding Basic Statistics*, confidence intervals for μ using large samples are presented. In Section 9.2, Student's *t* distribution is introduced and confidence intervals for μ using small samples are discussed. If the sample size n is small ($n < 30$) then the $\bar{x}$ distribution follows the Student's *t* distribution with degrees of freedom $(n - 1)$.

$$t = \frac{\bar{x} - \mu}{s/\sqrt{n}}$$

There is a different Student's *t* distribution for every degree of freedom. MINITAB includes Student's *t* distribution in its library of probability distributions. You may use the RANDOM, PDF, CDF, INVCDF commands with Student's *t* distribution as the specified distribution.

Menu selection: **Calc➤Probability Distribution➤t**

Dialog Box Responses

Select from Prob Density (PDF), Cumulative Probability (CDF), Inverse Cumulative Probability (INVCDF)

Degrees of Freedom: enter value

Input Column:

Column containing values for which you wish to compute the probability and optional storage column

Input Constant:

If you want the probability of just one value, use a constant rather than an entire column. Designate optional storage constant or column.

You can graph different *t*-distributions by using ➤**Graph**➤**Plot.** Follow steps similar to those given in Chapter 7 for graphing a normal distribution. Student's *t* distribution is symmetric and centered at 0. Select X values from about –4 to 4 in increments of 0.10 and place the values in a column, say C1. Then use **Calc➤Probability Distributions➤t** with Probability Density to generate a column, say C2, of Y values. The graph shown represents 10 degrees of freedom.

Student's Distribution with 10 Degrees of Freedom

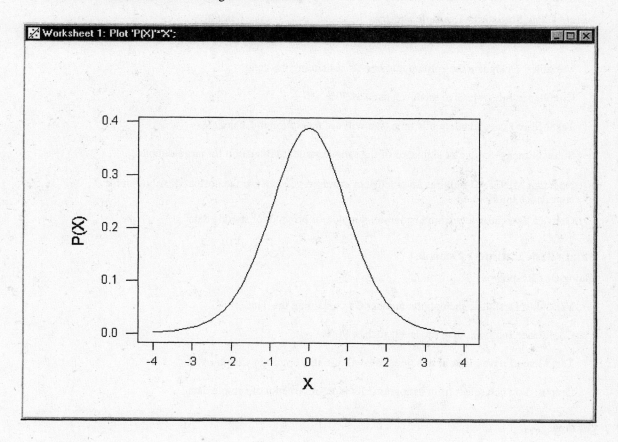

Confidence Intervals for Means

Confidence intervals for μ depend on the sample size n, and knowledge about the standard deviation σ. For small samples we assume the x distribution is approximately normal (mound-shaped and symmetric). The relationship between confidence intervals for μ, sample size, and knowledge about σ are shown in the table below.

Confidence Interval

Large samples

    σ known

    σ estimated by s $\bar{x} - z\left(\dfrac{\sigma}{\sqrt{n}}\right)$ to $\bar{x} + z\left(\dfrac{\sigma}{\sqrt{n}}\right)$

Small samples

    σ known

Small samples $\bar{x} - t\left(\dfrac{s}{\sqrt{n}}\right)$ to $\bar{x} + t\left(\dfrac{s}{\sqrt{n}}\right)$

    σ unknown

In MINITAB we can generate confidence intervals for μ by using the menu selections.

➤**Stat**➤**Basic Statistics**➤**1 sample z**

Dialog Box Responses

Variables: Designate the column number C# containing the data.

Confidence Interval: Give the level, such as 90%.

Test Mean: Leave blank at this time. We will use the option in Chapter 9.

Sigma: Enter the value of sigma, or of the sample standard deviation for large samples.

Note that MINITAB requires knowledge of σ before you can use the normal distribution for confidence intervals.

Graphs: You can select from histogram, dot plot, or box plot of sample data.

➤**Stat**➤**Basic Statistics**➤**1 sample t**

Dialog Box Responses

Variables: Designate the column number C# containing the data.

Confidence Interval: Give the level, such as 90%.

Test Mean: Leave blank at this time. We will use this option in Chapter 9.

Graphs: You can select from histogram, dot plot, or box plot of sample data.

Example

The manager of First National Bank wishes to know the average waiting times for student loan application action. A random sample of 20 applications showed the waiting times from application submission (in days) to be

3	7	8	24	6	9	12	25	18	17
4	32	15	16	21	14	12	5	18	16

Find a 90% confidence interval for the population mean of waiting times.

In this example we have a small sample and σ is not known. We need to use Student's t distribution. Enter the data into column C1 and name the column Days. Use the menu selection ➤**Stat**➤**Basic Statistics**➤**1 sample t.**

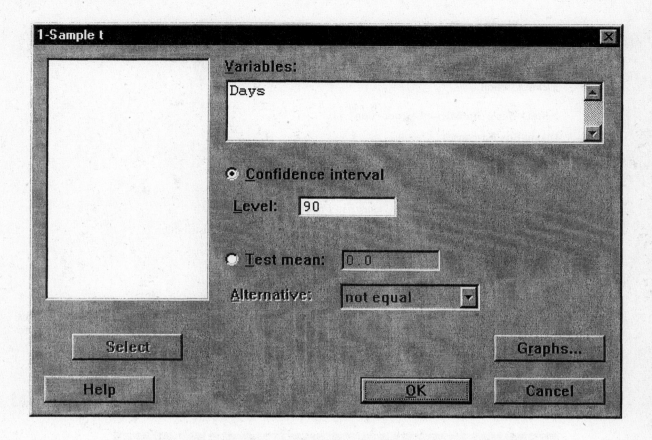

The results are

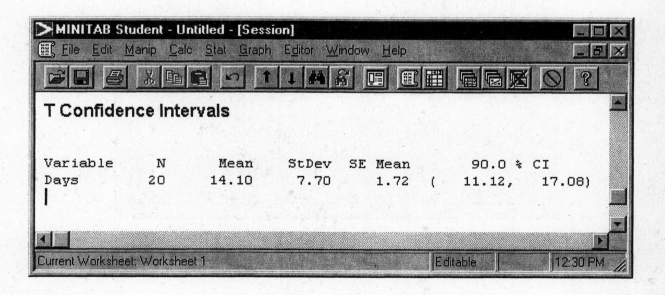

Confidence Intervals for Proportions

This option is in more recent versions of MINITAB, and is in Version 12.

Menu selection

➤Stat➤Basic Statistics➤1-proportion

Dialog Box Responses

Select the option of Summarized Data.

> Number of Trials: Enter value (*n* in *Understandable Statistics*)
>
> Number of Successes: Enter value (*r* in *Understandable Statistics*)

Click on [Options]; enter confidence level and click on Use normal distribution.

> In Chapter 10 we will see how to interpret the results from this test.

Example

The public television station BPBS wants to find the percent of its viewing population who give donations to the station. A random sample of 300 viewers were surveyed and it was found that 123 made contributions to the station. Find a 95% confidence interval for the probability that a viewer of BPBS selected at random contributions to the station.

Use the menu selection **➤Stat➤Basic Statistics➤1-proportion.** Click on Summarized Data. Use 300 for number of trials and 123 for number of successes. Click on [Options]. Enter 95 for the confidence level.

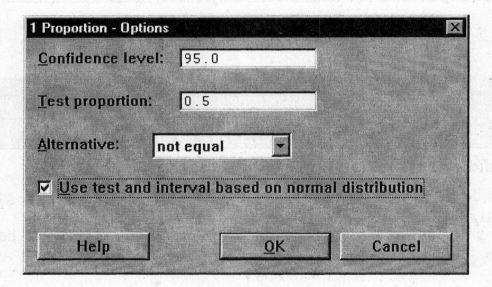

The results follow.

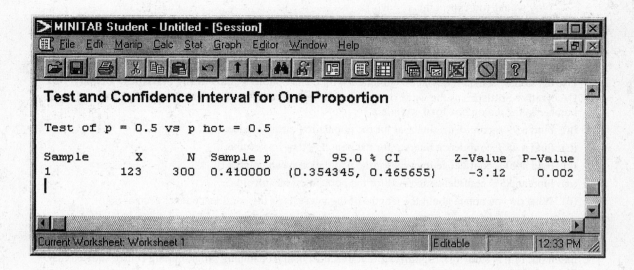

The output regarding test of p, Z-value, and P-value will be discussed in Chapter 10.

Confidence Intervals for Difference of Means or Difference of Proportions

In MINITAB, confidence intervals for difference of means and difference of proportions are included in the menu selection for tests of hypothesis for difference of means and tests of hypothesis for difference of proportions respectively. These menu selections with their dialog boxes will be discussed in Chapter 11.

LAB ACTIVITIES FOR CONFIDENCE INTERVALS FOR A MEAN OR FOR A PROPORTION

1. Snow King Ski resorts is considering opening a downhill ski slope in Montana. To determine if there would be an adequate snow base in November in the particular region under consideration, they studied snowfall records for the area over the last 100 years. They took a random sample of 15 years. The snowfall during November for the sample years was (in inches)

   26 35 42 18 29 42 28 35
   47 29 38 27 21 35 30

   (a) To find a confidence interval for μ, do we use a normal distribution or Student's t distribution?
   (b) Find a 90% confidence interval for the mean snowfall.
   (c) Find a 95% confidence interval for the mean snowfall.
   (d) Compare the intervals of parts (b) and (c). Which one is narrower? Why would you expect this?

2. Consider the snowfall data of problem 1. Suppose you knew that the snowfall in the region under consideration for the ski area in Montana (see problem 1) had a population standard deviation of 8 inches.
   (a) Since you know σ, (and the distribution of snowfall is assumed to be approximately normal) do you use the normal distribution or Student's t for confidence intervals?
   (b) Find a 90% confidence interval for the mean snowfall.
   (c) Find a 95% confidence interval for the mean snowfall.

(d) Compare the respective confidence intervals created in problem 1 and in this problem. Of the 95% intervals, which is longer, the one using the *t* distribution or the one using the normal distribution? Why would you expect this result?

3. Retrieve the worksheet Svls01.mpt from the CD-ROM. This worksheet contains the number of shares of Disney Stock (in hundreds of shares) sold for a random sample of 60 trading days in 1993 and 1994. The data is in column C1.

   Use the sample standard deviation computed with menu options ➤**Stat**➤**Basic Statistics**➤**Display Descriptive Statistics** as the value of σ. You will need to compute this value first, and then enter it as a number in the dialog box for 1-sample z.

   (a) Find a 99% confidence interval for the population mean volume.

   (b) Find a 95% confidence interval for the population mean volume.

   (c) Find a 90% confidence interval for the population mean volume.

   (d) Find an 85% confidence interval for the population mean volume.

   (d) What do you notice about the lengths of the intervals as the confidence level decreases?

4. There are many types of errors that will cause a computer program to terminate or give incorrect results. One type of error is punctuation. For instance, if a comma is inserted in the wrong place, the program might not run. A study of programs written by students in a beginning programming course showed that 75 out of 300 errors selected at random were punctuation errors. Find a 99% confidence interval for the proportion of errors made by beginning programming students that are punctuation errors. Next find a 90% confidence interval. Is this interval longer or shorter?

5. Sam decided to do a statistics project to determine a 90% confidence interval for the probability that a student at West Plains College eats lunch in the school cafeteria. He surveyed a random sample of 12 students and found that 9 ate lunch in the cafeteria. Can Sam use the program to find a confidence interval for the population proportion of students eating in the cafeteria? Why or why not? Try the program with N = 12 and R = 9. What happens? What should Sam do to complete his project?

COMMAND SUMMARY

Probability Distribution Subcommand

T with df = K is the subcommand that calls up Student's *t* distribution with specified degrees of freedom K. This subcommand may be used with RANDOM, PDF, CDF, INVCDF.

   WINDOWS menu selection: ➤**Calc**➤**Probability Distribution**➤**t**

   In the dialog box select PDF, CDF, or Inverse, then enter the degrees of freedom.

To Generate Confidence Intervals

ZINTERVAL [K% confidence] σ = K on C...C

   generates a confidence interval for μ using the normal distribution. You must enter a value for σ, either actual or estimated. A separate interval is given for data in each column. If K is not specified, a 95% confidence interval will be given.

   WINDOWS menu selection: ➤**Stat**➤**Basic Statistics**➤**1 sample z**

   In the dialog box select confidence interval and enter the confidence level.

TINTERVAL [K% confidence] for C...C

generates a confidence interval for μ using Student's t distribution. It automatically computes stdev, or s, from the data, as well as the number of degrees of freedom. If K is not specified, a 95% confidence interval is given.

WINDOWS menu selection: **➤Stat➤Basic Statistics➤1 sample t**

In the dialog box select Confidence interval and enter the confidence level.

PONE [N R] with **subcommand Confidence C** generates a confidence interval for one proportion.

WINDOWS menu selection: **➤Stat➤Basic Statistics➤1 proportion**

CHAPTER 10 HYPOTHESIS TESTING

TESTING A SINGLE POPULATION MEAN OR PROPORTION
(SECTIONS 10.1–10.5 OF *UNDERSTANDING BASIC STATISTICS*)

Tests involving a single mean are found in Sections 10.2 (large samples) and 10.4 (small samples). In MINITAB, the user concludes the test by comparing the P value of the test statistic to the level of significance α. The method of using P values to conclude tests of hypotheses is explained in Section 10.3. Section 10.5 discusses tests of a single proportion.

For tests of the mean with large samples use ➤**Stat**➤**Basic Statistics**➤**1-sample z.**

Dialog Box Responses

   Variable: Enter column number where data is located.

   Select Test Mean. Enter the value of k for the null hypothesis.

   $H_0: \mu = k$

   Alternative: Scroll to the appropriate alternate hypothesis:

   $H_1: \mu \neq k$ (not equal)

   $H_1: \mu > k$ (greater than)

   $H_1: \mu < k$ (less than)

   Sigma: Use the value of sigma, or estimate it with the sample standard deviation s.

   Recall that you can use the ➤**Stat**➤**Basic Statistics**➤**Display Descriptive Statistics** menu choices to find the value of s.

For tests of the mean with small samples use ➤**Stat**➤**Basic Statistics**➤**1-sample t.**

Dialog Box Responses

   Variable: Enter column number where data is located.

   Select Test Mean. Enter the value of k for the null hypothesis

   $H_0: \mu = k$

   Alternative: Scroll to the appropriate alternate hypothesis:

   $H_1: \mu \neq k$ (not equal)

   $H_1: \mu > k$ (greater than)

   $H_1: \mu < k$ (less than)

For tests of a single proportion use ➤**Stat**➤**Basic Statistics**➤**1 proportion.**

Dialog Box Responses

   Select Summarized data; enter the number of trials and the number of successes.

   Click on [Options].

   Confidence Level: Enter a value such as 95.

Test proportion: Enter the value of k, where

$$H_0: p = k$$

Alternative: Scroll to the appropriate alternate hypothesis:

$$H_1: p \neq k \text{ (not equal)}$$

$$H_1: p > k \text{ (greater than)}$$

$$H_1: p < k \text{ (less than)}$$

Both the Z-sample and the T-sample operate on data in a column. They each compute the sample mean $\overline{x}$. The Z-sample converts the sample mean $\overline{x}$ to a z value, while the T-sample converts $\overline{x}$ to t using the respective formulas

$$z = \frac{\overline{x} - \mu}{\sigma / \sqrt{n}} \qquad t = \frac{\overline{x} - \mu}{s / \sqrt{n}}$$

The test of 1 proportion converts the sample proportion $\hat{p} = r/n$ to a z value using the formula

$$z = \frac{\hat{p} - p}{\sqrt{p(1-p)/n}}$$

The tests also give the P value of the sample statistic $\overline{x}$. The user can then compare the P value to α, the level of significance of the test. If

P value $\leq \alpha$ we reject the null hypothesis.

P value $> \alpha$ we do not reject the null hypothesis.

Example

Many times patients visit a health clinic because they are ill. A random sample of 12 patients visiting a health clinic had temperatures (in °F) as follows:

97.4	99.3	99.0	100.0	98.6
97.1	100.2	98.9	100.2	98.5
98.8	97.3			

Dr. Tafoya believes that patients visiting a health clinic have a higher temperature than normal. The normal temperature is 98.6 degrees. Test the claim at the $\alpha = 0.01$ level of significance.

In this case, we have a small sample and do not know σ. We need a t-test. Enter the data in C1 and name the column Temp. Then select **➤Stat➤Basic Statistics➤1-sample t.**

Use a 98.6 as the value for Test mean. Scroll to greater than for Alternative.

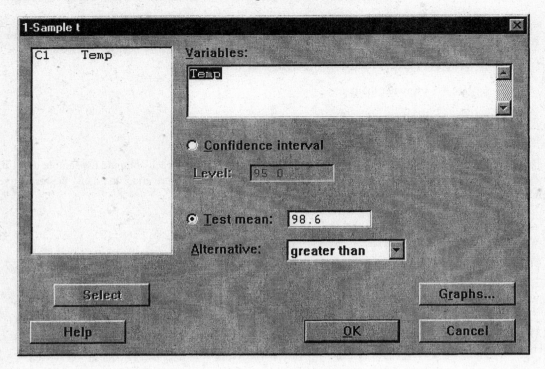

The results follow.

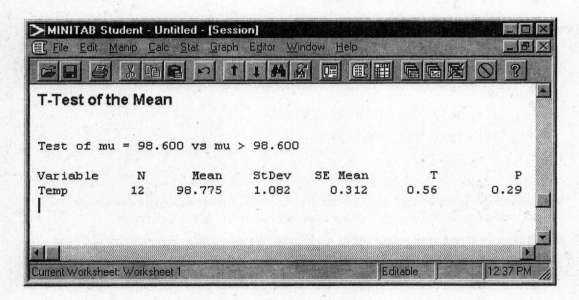

Recall that SE Mean is the value of $\dfrac{s}{\sqrt{n}}$.

LAB ACTIVITIES FOR TESTING A SINGLE POPULATION MEAN OR PROPORTION

1. A new catch-and-release policy was established for a river in Pennsylvania. Prior to the new policy, the average number of fish caught per fisherman hour was 2.8. Two years after the policy went into effect, a random sample of 12 fisherman hours showed the following catches per hour.

3.2	1.1	4.6	3.2	2.3	2.5
1.6	2.2	3.7	2.6	3.1	3.4

   Test the claim that the per hour catch has increased, at the 0.05 level of significance.

   (a) Decide whether to use the Z-sample or T-sample menu choices. What is the value of μ in the null hypothesis?

   (b) What is the choice ALTERNATIVE?

   (c) Compare the P value of the test statistic to the level of significance α. Do we reject the null hypothesis or not?

2. Open or retrieve the worksheet **Svls04.mtp** from the CD-ROM. The data in column C1 of this worksheet represent the miles per gallon gasoline consumption (highway) for a random sample of 55 makes and models of passenger cars (source: Environmental Protection Agency).

30	27	22	25	24	25	24	15
35	35	33	52	49	10	27	18
20	23	24	25	30	24	24	24
18	20	25	27	24	32	29	27
24	27	26	25	24	28	33	30
13	13	21	28	37	35	32	33
29	31	28	28	25	29	31	

   Test the hypothesis that the population mean miles per gallon gasoline consumption for such cars is greater than 25 mpg.

   (a) Do we know σ for the mpg consumption? Can we estimate σ by s, the sample standard deviation? Should we use the Z-sample or T-sample menu choice? What is the value of μ in the null hypothesis?

   (b) If we estimate σ by s, we need to instruct MINITAB to find the stdev, or s, of the data before we use Z-sample. Use ➤**Stat**➤**Basic Statistics**➤**Display Descriptive Statistics** to find s.

   (c) What is the alternative hypothesis?

   (d) Look at the P value in the output. Compare it to α. Do we reject the null hypothesis or not?

   (e) Using the same data, test the claim that the average mpg for these cars is not equal to 25. How has the P value changed? Compare the new P value to α. Do we reject the null hypothesis or not?

3. Open or retrieve the worksheet **Svss01.mtp** from the CD-ROM. The data in column C1 of this worksheet represent the number of wolf pups per den from a sample of 16 wolf dens (source: *The Wolf in the Southwest: The Making of an Endangered Species* by D.E. Brown, University of Arizona Press).

5	8	7	5	3	4	3	9
5	8	5	6	5	6	4	7

   Test the claim that the population mean number of wolf pups in a den is greater than 5.4.

4. Jones Computer Security is testing a new security device which is believed to decrease the incidence of computer break-ins. Without this device, the computer security test team can break security 47% of the time. With the device in place, the test team made 400 attempts and were successful 82 times. Select an appropriate program from the HYPOTHESIS TESTING menu and test the claim that the device reduces the proportion of successful break-ins. Use alpha = 0.05 and note the *P* value. Does the test conclusion change for alpha = 0.01?

COMMAND SUMMARY

<u>To Test a Single Mean</u>

ZTEST [μ = K] σ = K, for C…C performs a *z*-test on the data in each column. If you do not specify μ, it is assumed to be 0. You need to supply a value for σ (either actual, or estimated by the sample standard deviation *s* of a column in the case of large samples). If the ALTERNATIVE subcommand is not used, a two-tailed test is conducted.

> WINDOWS menu selection: ➤**Stat**➤**Basic Statistics**➤**1-sample z**

> In dialog box select alternate hypothesis, specify the mean for H_0, specify the standard deviation.

TTEST [μ = K] on C…C performs a separate *t*-test on the data of each column. If you do not specify μ, it is assumed to be 0. The computer evaluates *s*, the sample standard deviation for each column, and uses the computed *s* value to conduct the test. If the ALTERNATIVE subcommand is not used, a two-tailed test is conducted.

> WINDOWS menu selection: ➤**Stat**➤**Basic Statistics**➤**1-sample t**

> In dialog box select alternate hypothesis, specify the mean for H_0.

> **ALTERNATIVE = K** is the subcommand required to conduct a one-tailed test.

> If K = −1, then a left-tailed test is done. If K = 1, then a right-tailed test is done.

CHAPTER 11 INFERENCES ABOUT DIFFERENCES

TESTS INVOLVING PAIRED DIFFERENCES (DEPENDENT SAMPLES) (SECTION 11.1 OF *UNDERSTANDING BASIC STATISTICS*)

To perform a paired difference test, we put our paired data into two columns. New in later versions of MINITAB, including the Student Edition 12 for Windows, is a menu item for testing data pairs. Select

➤**Stat**➤**Basic Statistics**➤**Paired t**

Dialog Box Responses

First Sample: column number where before data is located

Second Sample: column number where after data is located

Click [Options]

Confidence Level: Enter a value such as 95.

Test mean: Leave as default 0.0.

Alternative: Scroll to not equal, greater than, or less than as appropriate.

H_1: $\mu \neq 0$ (not equal)

H_1: $\mu > 0$ (greater than)

H_1: $\mu < 0$ (less than)

Example

Promoters of a state lottery decided to advertise the lottery heavily on television for one week during the middle of one of the lottery games. To see if the advertising improved ticket sales, the promoters surveyed a random sample of 8 ticket outlets and recorded weekly sales for one week before the television campaign and for one week after the campaign. The results follow (in ticket sales) where B stands for "before" and A for "after" the advertising campaign.

B: 3201 4529 1425 1272 1784 1733 2563 3129

A: 3762 4851 1202 1131 2172 1802 2492 3151

We want to test to see if D = B – A is less than zero, since we are testing the claim that the lottery ticket sales are greater after the television campaign. We will put the before data in C1, the after data in C2. Select ➤**Stat**➤**Basic Statistics**➤**Paired t.** Use less than for Alternative, and use a Confidence level of 95.0.

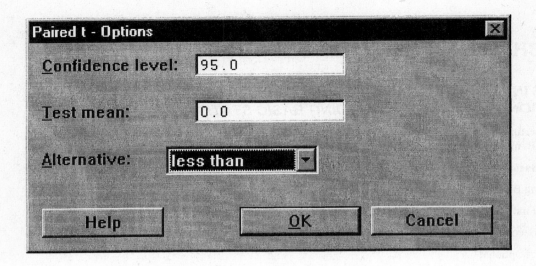

The results follow.

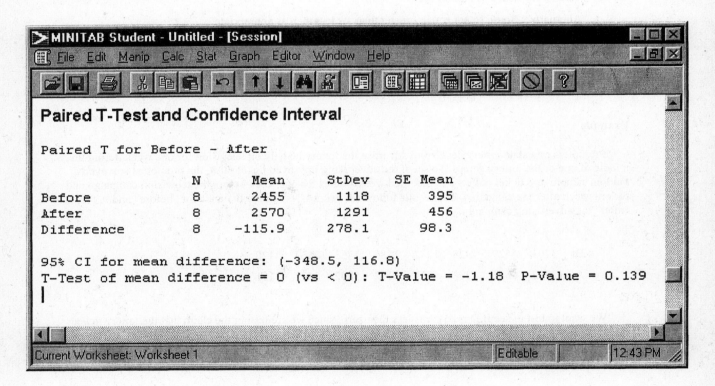

Since the *P* value 0.139 is larger than the level of significance of 0.05, we do not reject the null hypothesis.

LAB ACTIVITIES FOR TESTS INVOLVING PAIRED DIFFERENCES (DEPENDENT SAMPLES)

1. Open or retrieve the worksheet **Tvds01.mtp** from the CD-ROM. The data are pairs of values, where the entry in C1 represents the average salary ($1000/yr) for male faculty members at an institution and C2 represents the average salary for female faculty members ($1000/yr) at the same institution. A random sample of 22 U.S. colleges and universities was used (source: *Academe, Bulletin of the American Association of University Professors*).

   (34.5, 33.9)   (30.5, 31.2)   (35.1, 35.0)   (35.7, 34.2)   (31.5, 32.4)

   (34.4, 34.1)   (32.1, 32.7)   (30.7, 29.9)   (33.7, 31.2)   (35.3, 35.5)

   (30.7, 30.2)   (34.2, 34.8)   (39.6, 38.7)   (30.5, 30.0)   (33.8, 33.8)

   (31.7, 32.4)   (32.8, 31.7)   (38.5, 38.9)   (40.5, 41.2)   (25.3, 25.5)

   (28.6, 28.0)   (35.8, 35.1)

   (a) The data is in C1 and C2.

   (b) Use the ➤**Stat**➤**Basic Statistics**➤**Paired t** menu to test the hypothesis that there is a difference in salaries. What is the P value of the sample test statistic? Do we reject or fail to reject the null hypothesis at the 5% level of significance? What about at the 1% level of significance?

   (c) Use the ➤**Stat**➤**Basic Statistics**➤**Paired t** menu to test the hypothesis that female faculty members have a lower average salary than male faculty members. What is the test conclusion at the 5% level of significance? At the 1% level of significance?

2. An audiologist is conducting a study on noise and stress. Twelve subjects selected at random were given a stress test in a room that was quiet. Then the same subjects were given another stress test, this time in a room with high-pitched background noise. The results of the stress tests were scores 1 through 20, with 20 indicating the greatest stress. The results follow, where B represents the score of the test administered in the quiet room and A represents the scores of the test administered in the room with the high-pitched background noise.

Subject	1	2	4	5	6	7	8	9	10	11	12
B	13	12	16	19	7	13	9	15	17	6	14
A	18	15	14	18	10	12	11	14	17	8	16

Test the hypothesis that the stress level was greater during exposure to noise. Look at the P value. Should you reject the null hypothesis at the 1% level of significance? At the 5% level?

INFERENCES FOR DIFFERENCE OF MEANS (INDEPENDENT SAMPLES) (SECTION 11.2 OF *UNDERSTANDING BASIC STATISTICS*)

We consider the $\bar{x}_1 - \bar{x}_2$ distribution. The null hypothesis is that there is no difference between means, so $H_0: \mu_1 = \mu_2$, or $H_0: \mu_1 - \mu_2 = 0$.

Large Samples

MINITAB has a slightly different approach to testing difference of means with large samples (each sample size 30 or more) than that shown in *Understandable Statistics*. In MINITAB Student's *t* distribution is used instead of the normal distribution. The degrees of freedom used by MINITAB for this application of the *t* distribution is at least as large as the smaller sample. Therefore, we have degrees of freedom at 30 or more. In such cases the normal and Student's *t* distribution give reasonably similar results. However, the results will not be exactly the same.

The menu choice MINITAB uses to test the difference of means is **➤Stat➤Basic Statistics➤2-sample t**. The null hypothesis is always $H_0: \mu_1 = \mu_2$. The alternate hypothesis $H_1: \mu_1 \neq \mu_2$, corresponds to the choice "not equal." To do a left-tailed or right-tailed test, you need to use the choice of "less than" for ALTERNATIVE on a left-tailed test and "greater than" for ALTERNATIVE on a right-tailed test.

WINDOWS menu selection: **➤Stat➤Basic Statistics➤2-sample t**

Dialog Box Responses

Select Samples in Different Columns and enter the C# for the columns containing the data.

Alternative: Scroll to the appropriate choice.

Confidence Level: Enter a value such as 95.

Assume equal variances: Do **not** select for large samples.

Small Samples

To do a test of difference of sample means with small samples with the assumption that the samples come from populations with the same standard deviation, we use the **➤Stat➤Basic Statistics➤2-sample t** menu selection with **Assume equal variances** checked. When we check that equal variances are assumed, MINITAB automatically pools the standard deviations.

➤Stat➤Basic Statistics➤2-sample t

Dialog Box Responses

Select Samples in Different Columns and enter the C# for the columns containing the data.

Alternative: Scroll to the appropriate choice.

Confidence Level: Enter a value such as 95.

Assume equal variances: Check this item so that the pooled standard deviation is used.

Example

Sellers of microwave French fry cookers claim that their process saves cooking time. McDougle Fast Food Chain is considering the purchase of these new cookers, but wants to test the claim. Six batches of French fries were cooked in the traditional way. Cooking times (in minutes) are

15   17   14   15   16   13

Six batches of French fries of the same weight were cooked using the new microwave cooker. These cooking times (in minutes) are

11   14   12   10   11   15

Test the claim that the microwave process takes less time. Use $\alpha = 0.05$.

Under the assumption that the distributions of cooking times for both methods are approximately normal and that $\sigma_1 = \sigma_2$, we use the **➤Stat➤Basic Statistics➤2-sample t** menu choices with the assumption of equal variances checked. We are testing the claim that the mean cooking time of the second sample is less than that of the first sample, so our alternate hypothesis will be $H_1: \mu_1 > \mu_2$. We will use a right-tailed test and scroll to "greater than" for ALTERNATIVE.

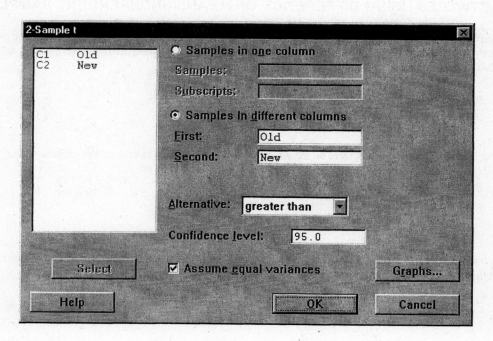

The results follow.

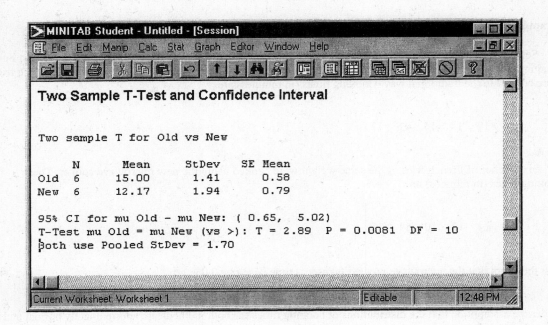

We see that the *P* value of the test is 0.0081. Since the *P* value is less than $\alpha = 0.05$, we reject the null hypothesis and conclude that the microwave method takes less time to cook French fries.

LAB ACTIVITIES USING DIFFERENCE OF MEANS (INDEPENDENT SAMPLES)

1. Calm Cough Medicine is testing a new ingredient to see if its addition will lengthen the effective cough relief time of a single dose. A random sample of 15 doses of the standard medicine were tested, and the effective relief times were (in minutes):

42	35	40	32	30	26	51	39	33	28
37	22	36	33	41					

   A random sample of 20 doses were tested when the new ingredient was added. The effective relief times were (in minutes):

43	51	35	49	32	29	42	38	45	74
31	31	46	36	33	45	30	32	41	25

   Assume that the standard deviations of the relief times are equal for the two populations. Test the claim that the effective relief time is longer when the new ingredient is added. Use $\alpha = 0.01$.

2. Open or retrieve the worksheet **Tvis06.mtp** from the CD-ROM. The data represent the number of cases of red fox rabies for a random sample of 16 areas in each of two different regions of southern Germany.

NUMBER OF CASES IN REGION 1

10 2 2 5 3 4 3 3 4 0 2 6 4 8 7 4

NUMBER OF CASES IN REGION 2

1 1 2 1 3 9 2 2 4 5 4 2 2 0 0 2

Test the hypothesis that the average number of cases in Region 1 is greater than the average number of cases in Region 2. Use a 1% level of significance.

3. Open or retrieve the MINITAB worksheet **Tvis02.mtp** from the CD-ROM. The data represent the petal length (cm) for a random sample of 35 Iris Virginica and for a random sample of 38 Iris Setosa (source: Anderson, E., Bull. Amer. Iris Soc).

PETAL LENGTH (CM) IRIS VIRGINICA

5.1	5.8	6.3	6.1	5.1	5.5	5.3	5.5	6.9	5.0	4.9	6.0	4.8	6.1	5.6	5.1
5.6	4.8	5.4	5.1	5.1	5.9	5.2	5.7	5.4	4.5	6.1	5.3	5.5	6.7	5.7	4.9
4.8	5.8	5.1													

PETAL LENGTH (CM) IRIS SETOSA

1.5	1.7	1.4	1.5	1.5	1.6	1.4	1.1	1.2	1.4	1.7	1.0	1.7	1.9	1.6	1.4
1.5	1.4	1.2	1.3	1.5	1.3	1.6	1.9	1.4	1.6	1.5	1.4	1.6	1.2	1.9	1.5
1.6	1.4	1.3	1.7	1.5	1.7										

Test the hypothesis that the average petal length for the Iris Setosa is shorter than the average petal length for the Iris Virginica.

COMMAND SUMMARY

Confidence Intervals and Tests of Paired Difference (Dependent Samples)

(Available on newer versions of MINITAB, such as release 12)

PAIRED C...C tests for a difference of means in paired (dependent) data and gives a confidence interval if requested.

TEST 0.0 is a subcommand to set the null hypothesis to 0

ALTERNATIVE = K is the subcommand to change the alternate hypothesis to a left-tailed test with K = –1 or right-tailed test with K = 1.

> WINDOWS menu selection: ➤**Stat**➤**Basic Statistics**➤**paired t**

> In dialog box select alternate hypothesis, specify the mean for H_0.

Confidence Intervals and Tests of Difference of Means (Independent Samples)

TWOSAMPLE [K% confidence] for C...C does a two (independent) sample *t* test and (optionally) confidence interval for data in the two columns listed. The first data set is put into the first column, and the second data set into the second column. Unless the ALTERNATIVE subcommand is used, the alternate hypothesis is assumed to be H_1: $\mu_1 \neq \mu_2$. Samples are assumed to be independent.

ALTERNATIVE = K is the subcommand to change the alternate hypothesis to a left-tailed test with K = –1 or right-tailed test with K = 1.

POOLED is the subcommand to be used only when the two samples come from populations with equal standard deviations.

> WINDOWS menu selection: ➤**Stat**➤**Basic Statistics**➤**2-sample t**

> In dialog box select alternate hypothesis, specify the mean for H_0. For large samples do not check assume equal variances. For small samples check assume equal variances.

CHAPTER 12 ADDITIONAL TOPICS USING INFERENCES

CHI-SQUARE TESTS OF INDEPENDENCE
(SECTION 12.1 OF *UNDERSTANDING BASIC STATISTICS*)

In chi-square tests of independence we use the hypotheses.

H_0: The variables are independent

H_1: The variables are not independent

To use MINITAB for tests of independence, we enter the values of a contingency table row by row. The command CHISQUARE then prints a contingency table showing both the observed and expected counts. It computes the sample chi-square value using the following formula, in which E stands for the expected count in a cell and O stands for the observed count in that same cell. The sum is taken over all cells.

$$\chi^2 = \sum \frac{(O-E)^2}{E}$$

Then MINITAB gives the number of degrees of the chi-square distribution. To conclude the test, use the P value of the sample chi-square statistic if your version of MINITAB provides it. Otherwise, compare the calculated chi-square value to a table of the chi-square distribution with the indicated degrees of freedom. We may use Table 6 of Appendix I of *Understanding Basic Statistics*. If the calculated sample chi-square value is larger than the value in Table 6 for a specified level of significance, we reject H_0.

Use the menu selection

➤**Stat**➤**Tables**➤**Chi-square Test**

Dialog Box Responses

List the columns containing the data from the contingency table. Note that you may use up to seven columns of data. Each column must contain integer values.

Example

A computer programming aptitude test has been developed for high school seniors. The test designers claim that scores on the test are independent of the type of school the student attends: rural, suburban, urban. A study involving a random sample of students from these types of institutions yielded the following contingency table. Use the CHISQUARE command to compute the sample chi-square value, and to determine the degrees of freedom of the chi-square distribution. Then determine if type or school and test score are independent at the $\alpha = 0.05$ level of significance.

School Type

Score	Rural	Suburban	Urban
200–299	33	65	83
300–399	45	79	95
400–500	21	47	63

To use the menu selection ➤**Stat**➤**Tables**➤**Chi-square Test** with C1 containing test scores for rural schools, C2 corresponding test scores for suburban schools, and C3 corresponding test scores for urban schools.

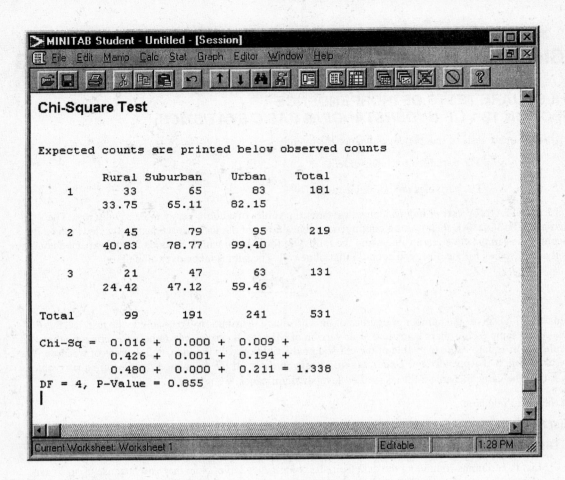

Since the *P* value, 0.855, is greater than $\alpha = 0.05$, we do not reject the null hypothesis.

LAB ACTIVITIES FOR CHI-SQUARE TESTS OF INDEPENDENCE

Use MINITAB to produce a contingency table and compute the sample chi square value. If your version of MINITAB produces the *P* value of the sample chi-square statistic, conclude the test using *P* values.

1. We Care Auto Insurance had its staff of actuaries conduct a study to see if vehicle type and loss claim are independent. A random sample of auto claims over six months gives the information in the contingency table.

Total Loss Claims per Year per Vehicle

Type of vehicle	$0–999	$1000–2999	$3000–5999	$6000+
Sports car	20	10	16	8
Truck	16	25	33	9
Family Sedan	40	68	17	7
Compact	52	73	48	12

Test the claim that car type and loss claim are independent. Use $\alpha = 0.05$.

2. An educational specialist is interested in comparing three methods of instruction.

SL–standard lecture with discussion

TV–video taped lectures with no discussion

IM–individualized method with reading assignments

and tutoring, but no lectures.

The specialist conducted a study of these methods to see if they are independent. A course was taught using each of the three methods and a standard final exam was given at the end. Students were put into the different method sections at random. The course type and test results are shown in the next contingency table.

Final Exam Score

Course Type	< 60	60–69	70–79	80–89	90–100
SL	10	4	70	31	25
TV	8	3	62	27	23
IM	7	2	58	25	22

Test the claim that the instruction method and final exam test scores are independent, using $\alpha = 0.01$.

SIMPLE LINEAR REGRESSION: TWO VARIABLES
(SECTIONS 12.4 AND 12.5 OF *UNDERSTANDING BASIC STATISTICS*)

Chapter 4 of *Understanding Basic Statistics* introduces linear regression. Formulas to find the equation of the least-squares line

$$y = a + bx$$

are given in Section 4.2 and reviewed in Section 12.4. Section 12.4 also contains the equation for the standard error of estimate as well as the procedure to find a confidence interval for the predicted value of y. The formula for the correlation coefficient r and coefficient of determination r^2 are given in Section 4.3. Sections 12.4 and 12.5 continue this discussion with inferences relating to linear regression. In Section 12.5, we use the sample correlation to test the population correlation coefficient ρ .

The menu selection ➤**Stat**➤**Regression**➤**Regression** gives the equation of the least-squares line, the value of the standard error of estimate (s = standard error of estimate), the value of the coefficient of determination r^2 (R-sq), as well as several other values such as R – sq adjusted (an unbiased estimate of the population r^2). For simple regression with a response variable and one explanatory variable, we can get the value of the Pearson product moment correlation coefficient r by simply taking the square root of R-sq.

The standard deviation, t-ratio and P values of the coefficients are also given.

Depending on the amount of output requested (controlled by the options selected under the [**Results**] button) you will also see an analysis of variance chart, as well as a table of x and y values with the fitted values y_p and residuals $(y - y_r)$. We will not use the analysis of variance chart in our introduction to regression. However, in more advanced treatments of regression, you will find it useful.

To find the equation of the least-squares line and the value of the correlation coefficient, use the menu options

➤Stat➤Regression➤Regression

Dialog Box Responses

Response: Enter the column number C# of the column containing the responses (that is Y values).

Predictor: Enter the column number C# of the column containing the explanatory variables (that

is, X values).

[Graphs]: Do not click on at this time.

[Results]: Click on and select the second option, Regression equation, etc.

[Options]: We will click on this option when we wish to do predictions for new variables.

[Storage]: Click on and select the fits and residual options if you wish.

To graph the scatter plot and show the least-squares line on the graph, use the menu options

➤Stat➤Regression➤Regression

Dialog Box Responses

Response: List the column number C# of the column containing the Y values.

Predictor: List the column number C# of the column containing the X values.

Type of Regression model: Select Linear.

[Options]: Click on and select Display Prediction Band for a specified confidence level of

prediction band. Do not use if you do not want the prediction band.

[Storage]: This button gives you the same storage options as found under regression.

To find the value of the correlation coefficient directly and to find its corresponding P value, use the menu selection

➤Stat➤Regression➤Regression

Dialog Box Responses

Variables: List the column number C# of the column containing the X variable and the column

number C# of the column containing the Y variable.

Select the P value option.

Example

Merchandise loss due to shoplifting, damage, and other causes is called shrinkage. Shrinkage is a major concern to retailers. The managers of H.R. Merchandise think there is a relationship between shrinkage and number of clerks on duty. To explore this relationship, a random sample of 7 weeks was selected. During each week the staffing level of sales clerks was kept constant and the dollar value (in hundreds of dollars) of the shrinkage was recorded.

X	10	12	11	15	9	13	8	
Y	19	15	20	9	25	12	31	(in hundreds)

Store the value of X in C1 and name C1 as X. Sore the values of Y in C2 and name C2 as Y.

Use menu choices to give descriptive statistics regarding the values of X and Y. Use commands to draw an (X, Y) scatter plot and then to find the equation of the regression line. Find the value of the correlation coefficient, and test to see if it is significant.

(a) First we will use ➤**Stat**➤**Basic Statistics**➤**Display Descriptive Statistics** and each of the columns X, and Y. Note that we select both C1 and C2 in the variables box.

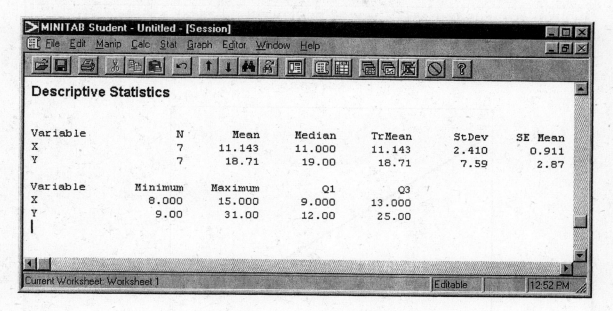

```
Descriptive Statistics


Variable          N       Mean     Median     TrMean      StDev     SE Mean
X                 7     11.143     11.000     11.143      2.410       0.911
Y                 7      18.71      19.00      18.71       7.59        2.87

Variable    Minimum    Maximum         Q1         Q3
X             8.000     15.000      9.000     13.000
Y              9.00      31.00      12.00      25.00
```

(b) Next we will use ➤**Stat**➤**Regression**➤**Fitted Line Plot** to graph the scatter plot and show the least-squares line on the graph. We will not use prediction bands.

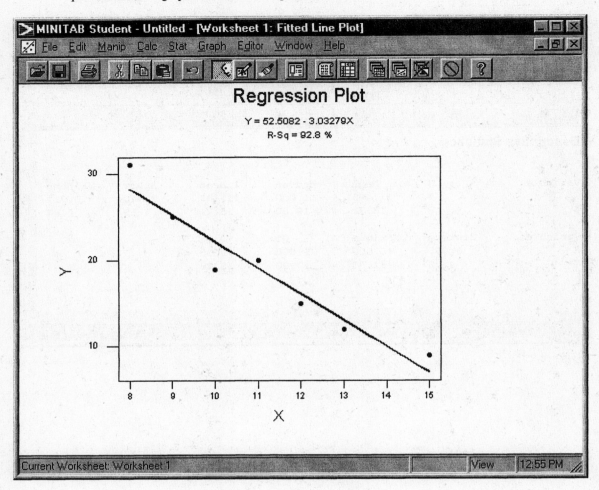

Notice that the equation of the regression line is given on the figure, as well as the value of r^2.

(c) However, to find out more information about the linear regression model, we use the menu selection
➤**Stat**➤**Regression**➤**Regression.** Enter C2 for Response and C1 for Predictor.

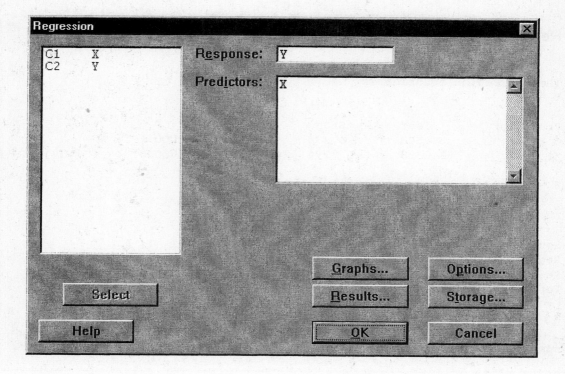

The results follow.

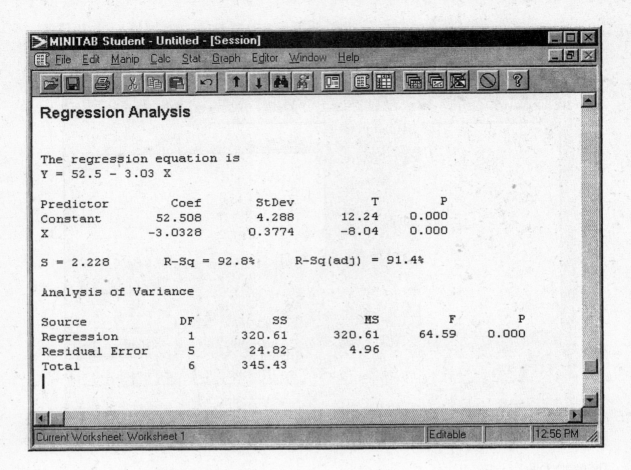

Notice that the regression equation is given as

$$y = 52.5 - 3.03x$$

The value of the standard error of estimate S_e is given as $S = 2.28$. We have the value of r^2, R-square = 92.8%. Find the value of r by taking the square root. It is 0.963 or 96.3%.

(d) Next, let's use the prediction option to find the shrinkage when 14 clerks are available.

Use ➤ **Stat** ➤ **Regression** ➤ **Regression.** Your previous selections should still be listed. Now press [Options]. Enter 14 in the prediction window.

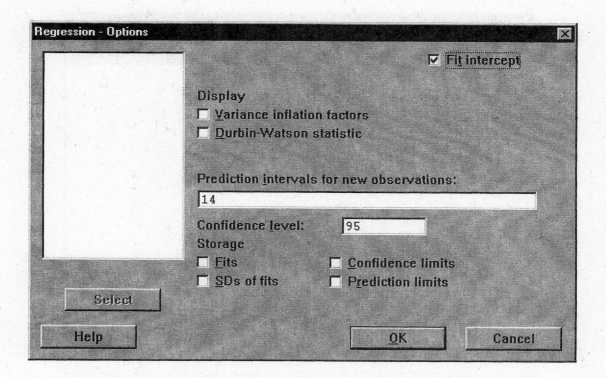

The results follow.

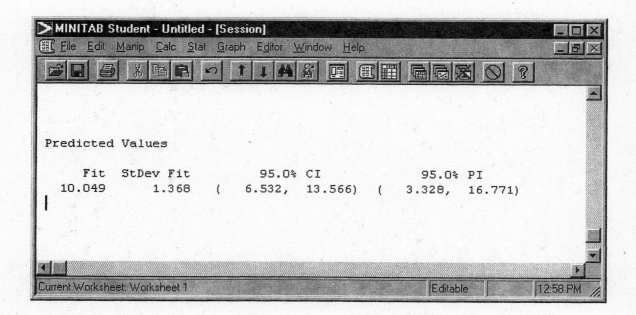

The predicted value of the shrinkage when 14 clerks are on duty is 10.05 hundred dollars, or $1005. A 95% prediction interval goes from 3.33 hundred dollars to 16.77 hundred dollars—that is, from $333 to $1677.

(e) Graph a prediction band for predicted values.

Now we use ➤**Stat**➤**Regression**➤**Fitted Line Plot** with the [Option] Display prediction bands selected.

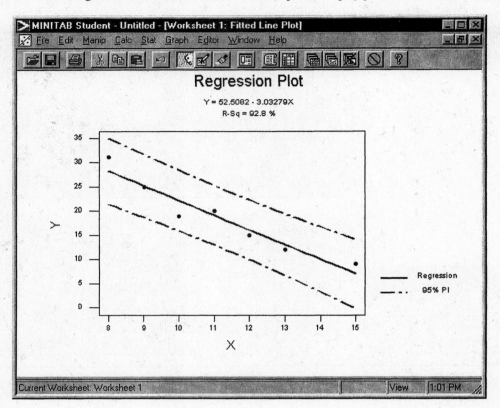

(f) Find the correlation coefficient and test it against the hypothesis that there is no correlation. We use the menu options **➤Stat➤Basic Statistics➤Correlation.**

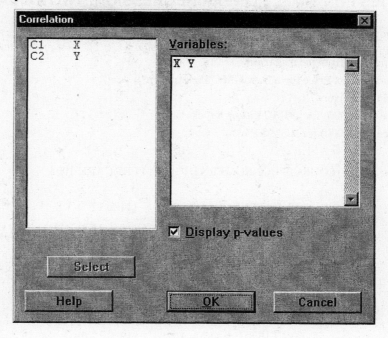

The results are

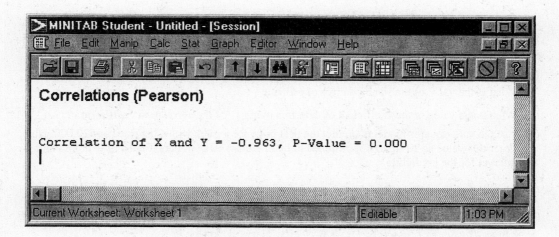

Notice $r = -0.963$ and the P value is so small as to be 0. We reject the null hypothesis and conclude that there is a linear correlation between the number of clerks on duty and the amount of shrinkage.

LAB ACTIVITIES FOR SIMPLE LINEAR REGRESSION: TWO VARIABLES

1. Open or retrieve the worksheet **Slr01.mtp** from the CD-ROM. This worksheet contains the following data, with the list price in column C1 and the best price in the column C2. The best price is the best price negotiated by a team from the magazine.

   List Price versus Best Price for a New GMC Pickup Truck

   In the following data pairs (X, Y)

   X = List Price (in $1000) for a GMC Pickup Truck

   Y = Best Price (in $1000) for a GMC Pickup Truck

   SOURCE: CONSUMERS DIGEST, FEBRUARY 1994

(12.4, 11.2)	(14.3, 12.5)	(14.5, 12.7)
(14.9, 13.1)	(16.1, 14.1)	(16.9, 14.8)
(16.5, 14.4)	(15.4, 13.4)	(17.0, 14.9)
(17.9, 15.6)	(18.8, 16.4)	(20.3, 17.7)
(22.4, 19.6)	(19.4, 16.9)	(15.5, 14.0)
(16.7, 14.6)	(17.3, 15.1)	(18.4, 16.1)
(19.2, 16.8)	(17.4, 15.2)	(19.5, 17.0)
(19.7, 17.2)	(21.2, (18.6)	

   (a) Use MINITAB to find the least-squares regression line using the best price as the response variable and list price as the explanatory variable.

   (b) Use MINITAB to draw a scatter plot of the data.

   (c) What is the value of the standard error of estimate?

   (d) What is the value of the coefficient of determination r^2? of the correlation coefficient r?

   (e) Use the least-squares model to predict the best price for a truck with a list price of $20,000. Note: Enter this value as 20 since X is assumed to be in thousands of dollars. Find a 95% confidence interval for the prediction.

2. Other MINITAB worksheets appropriate to use for simple linear regression are

   Cricket Chirps Versus Temperature: **Slr02.mtp**

   Source: *The Song of Insects* by Dr. G.W. Pierce, Harvard College Press

   The chirps per second for the striped grouped cricket are stored in C1; the corresponding temperature in degrees Fahrenheit is stored in C2.

   Diameter of Sand Granules Versus Slope on a Beach:

   Slr03.mtp; source *Physical Geography* by A.M. King, Oxford press

   The median diameter (mm) of granules of sand in stored in C1; the corresponding gradient of beach slope in degrees is stored in C2.

   National Unemployment Rate Male Versus Female: **Slr04.mtp**

   Source: *Statistical Abstract of the United States*

   The national unemployment rate for adult males is stored in C1; the corresponding unemployment rate for adult females for the same period of time is stored in C2.

The data in these worksheets are described in the Appendix of this *Guide*. Select these worksheets and repeat parts (a)–(d) of problem 1, using C1 as the explanatory variable and C2 as the response variable.

3. A psychologist interested in job stress is studying the possible correlation between interruptions and job stress. A clerical worker who is expected to type, answer the phone and do reception work has many interruptions. A store manager who has to help out in various departments as customers make demands also has interruptions. An accountant who is given tasks to accomplish each day and who is not expected to interact with other colleagues or customers except during specified meeting times has few interruptions. The psychologist rated a group of jobs for interruption level. The results follow, with X being interruption level of the job on a scale of 1 to 20, with 20 having the most interruptions, and Y the stress level on a scale of 1 to 50, with 50 the most stressed.

Person		1	2	3	4	5	6	7	8	9	10	11	12
	X	9	15	12	18	20	9	5	3	17	12	17	6
	Y	20	37	45	42	35	40	20	10	15	39	32	25

(a) Enter the X values into C1 and the Y values into C2. Use the menu selections ➤Stat➤Basic Statistics➤Display Descriptive Statistics on the two columns. What is the mean of the Y-values? Of the X-values? What are the respective standard deviations?

(b) Make a scatter plot of the data using the ➤Stat➤Regression➤Fitted Line menu selection. From the diagram do you expect a positive or negative correlation?

(c) Use the ➤Stat➤Basic Statistics➤Correlation menu choices to get the value of *r*. Is this value consistent with your response in part (b)?

(d) Use the ➤Stat➤Regression➤Regression menu choices with Y as the response variable and X as the explanatory variable. Use the [Option] button with predictions 5, 10, 15, 20 to get the predicted stress level of jobs with interruption levels of 5, 10, 15, 20. Look at the 95% P.I. intervals. Which are the longest? Why would you expect these results? Find the standard error of estimate. Is R-sq equal to the square of *r* as you found in part (c)? What is the equation of the least-squares line?

(e) Redo the ➤Stat➤Regression➤Regression menu option, this time using X as the response variable and Y as the explanatory variable. Is the equation different than that of part (d)? What about the value of the standard error of estimate (s on your output)? Did it change? Did R-sq change?

4. The researcher of problem 3 was able to add to her data. Another random sample of 11 people had their jobs rated for interruption level and were then evaluated for stress level.

Person		13	14	15	16	17	18	19	20	21	22	23
	X	4	15	19	13	10	9	3	11	12	15	4
	Y	20	35	42	37	40	23	15	32	28	38	12

Add this data to the data in problem 3, and repeat parts (a) through (e). Compare the values of s, the standard error of estimate in parts (d). Did more data tend to reduce the value of s? Look at the 95% P.I. intervals. How do they compare to the corresponding ones of problem 3? Are they shorter or longer? Why would you expect this result?

COMMAND SUMMARY

CHISQUARE test on table stored in C...C

produced a contingency table and computes the sample chi-square value

WINDOWS menu select: ➤**Stat**➤**Tables**➤**Chi-square test**

In the dialog box, specify the columns which contain the chi-square table.

To Perform Simple or Multiple Regression

REGRESS C on K explanatory variables in C...C

does regression with the first column containing the response variable, K explanatory variables in the remaining columns.

PREDICT E...E predicts the response variable for the given values of the explanatory variable(s).

RESIDUALS put into C stores the residuals in column C.

WINDOWS menu selection: ➤**Stat**➤**Regression**➤**Regression**

Use the dialog box to list the response and explanatory (prediction) variables. Mark the residuals box. In the Options dialog box list the values of the explanatory variable(s) for which you wish to make a prediction. Select the P.I. confidence interval.

BRIEF K controls the amount of output for K = 1, 2, 3 with 3 giving the most output.

This command is not available from a menu.

There are other subcommands for REGRESS. See the MINITAB Help for your release of MINITAB for a list of the subcommands and their descriptions.

To Find the Pearson Product Moment Correlation Coefficient

CORRELATION for C...C calculates the correlation coefficient for all pairs of columns.

WINDOWS menu selection: ➤**Stat**➤**Basic Statistics**➤**Correlation**

To Graph the Scatter Plot for Simple Regression

With **GSTD** use the **PLOT C vs C** command.

WINDOWS menu selection (MINITAB SE for Windows or MINITAB 10 for Windows):

➤**Stat**➤**Regression**➤**Fitted Line Plot**

COMMAND REFERENCE

This appendix summarizes all the MINITAB commands used in this Guide. A complete list of commands may be found in the MINITAB Reference Manual that comes with the MINITAB software.

> C denotes a column
>
> E denotes either a column or constant
>
> K denotes a constant
>
> [] denotes optional parts of the command

GENERAL INFORMATION

HELP   gives general information about MINITAB.

   WINDOWS menu: **Help**

STOP   ends MINITAB session

   WINDOWS menu: **➤File➤Exit**

TO ENTER DATA

READ C...C   puts data into designated columns.

READ 'filename' C...C reads data from file into columns.

SET C   put data into single designated column.

SET 'filename' C   reads data from file into column.

END   signals end of data.

NAME C = 'name' names column C.

   WINDOWS menu selection: You can enter data in rows or columns and name the column in the

   DATA window. To access the data window select **➤Window➤Data.**

RETRIEVE 'filename' retrieves worksheet.

   WINDOWS menu selection: **➤File➤Retrieve**

TO EDIT DATA

LET C(K) = K   changes the value in row K of column C.

INSERT K K C C inserts data between rows K and K into columns C to C.

DELETE K K C C deletes data between row K and K from columns C to C.

   WINDOWS menu selection: You can edit data in rows or columns in the DATA window.

   To access the data window select **➤Window➤Data.**

COPY C into C   copies column C into column C.

USE rows K…K   subcommand to copy designated rows

OMIT rows K…K   subcommand to omit designated rows

    WINDOWS menu selection: ➤**Manip**➤**Copy Columns**

ERASE E…E   erases designated columns or constraints.

    WINDOWS menu selection: ➤**Manip**➤**Erase Variables**

TO OUTPUT DATA

PRINT E…E   prints designated columns or constant.

    WINDOWS menu selection: ➤**File**➤**Display Data**

SAVE 'filename' saves current worksheet.

PORTABLE   subcommand to make worksheet portable

    WINDOWS menu selection: ➤**File**➤**Save Worksheet**

    WINDOWS menu selection: ➤**File**➤**Save Worksheet As…** you may select portable.

WRITE 'filename' C…C saves data in ASCII file.

    WINDOWS menu selection: ➤**File**➤**Other Files**➤**Export ASCII Data**

MISCELLANEOUS

PAPER   prints session.

   NOPAPER   stops printing session.

    WINDOWS menu selection: ➤**File**➤**Print Window**

OUTFILE = 'filename' saves session in ASCII file.

NOOUTFILE   ends OUTFILE.

    WINDOWS menu selection: ➤**File**➤**Other Files**➤**Start/Stop Recording**

TO DO ARITHMETIC

LET E = expression evaluates the expression and stores the result in E, where E may be a column or a constant.

   ** raise to a power

   * multiplication

   / division

   + addition

   − subtraction

SQRT(E)   takes the square root.

ROUND(E) rounds numbers to the nearest integer.

    There are other arithmetic operations possible.

    WINDOW menu selection: ➤**Calc**➤**Calculator**

TO GENERATE A RANDOM SAMPLE

RANDOM K into C…C selects a random sample from the distribution described in the subcommand.

> **INTEGER K to K**   distribution of integers from K to K
>
> **BERNOULLI P = K**
>
> **BINOMIAL N = K, P = K**
>
> **CHISQUARE degrees of freedom = K**
>
> **DISCRETE values in C probabilities in C**
>
> **F df numerator = K, df denominator = K**
>
> **NORMAL [mean = K [standard deviation = K]]**
>
> **POISSON mean = K**
>
> **T degrees of freedom = K**
>
> **UNIFORM continuous distribution on [K to K]**
>
> > WINDOWS menu selection: ➤**Calc**➤**Random data**➤**Select Distribution**

SAMPLE K rows from C…C and put results in C…C takes a random sample of rows without replacement.

REPLACE cause the sample to be taken with replacement.

TO ORGANIZE DATA

SORT C…C PUT IN C…C sorts the data in the first column and carries the other columns along.

DESCENDING C…C   subcommand to sort in descending order

> WINDOWS menu selection: ➤**Manip**➤**Sort**

TALLY data in C…C   tallies data in columns. The data must be integer values.

> **COUNTS**
>
> **PERCENTS**
>
> **CUMCOUNTS**
>
> **CUMPERCENTS**
>
> **ALL**   gives all four values.
>
> > WINDOWS menu selection: ➤**Stats**➤**Tables**➤**Tally**

HISTOGRAM C…C prints a separate histogram for data in each of the listed columns.

> **START with midpoint = k [end with endpoint = K]**
>
> **INCREMENT = K**   specifies distance between midpoints
>
> > WINDOWS menu selection: (for professional graphics) ➤**Graph**➤**Histogram (options for cutpoints)**
> >
> > WINDOWS menu selection: (for character graphics) ➤**Graph**➤**Character Graphs**➤**Histogram**

STEM-AND-LEAF DISPLAY OF C...C

makes separate stem-and-leaf displays of data in each of the listed columns.

INCREMENT = K sets the distance between two display lines.

TRIM outliers   lists extreme data on special lines.

WINDOWS menu selection: (for character graphics) ➤**Graph**➤**Character Graphs**➤**Stem-and-Leaf**

BOXPLOT C   makes a box-and-whisker plot of data in column C.

START = K [end = k]

INCREMENT = K

WINDOWS menu selection: (for character graphics) ➤**Graph**➤**Character Graphs**➤**Box**

WINDOWS menu selection: (for professional graphics) ➤**Graph**➤**Boxplot**

TO SUMMARIZE DATA BY COLUMN

DESCRIBE C...C prints descriptive statistics

WINDOWS menu selection: ➤**Stat**➤**Descriptive Statistics**

COUNT   C [**put into K**] counts the values.

N   C [**put into K**] counts the non-missing values.

NMIS   C [**put into K**] counts the missing values.

SUM   C [**put into K**] sums the values.

MEAN   C [**put into K**] gives arithmetic mean of values.

STDEV   C [**put into K**] gives standard deviation.

MEDIAN   C [**put into K**] gives the median of the values.

MINIMUM C [**put into K**] gives the minimum of the values.

MAXIMUM C [**put into K**] gives the maximum of the values.

SSQ   C [**put into K**] gives the sum of squares of values.

TO SUMMARIZE DATA BY ROW

RCOUNT   E...E put into C

RN   E...E put into C

RNMIS   E...E put into C

RSUM   E...E put into C

RMEAN   E...E put into C

RSTDEV   E...E put into C

RMEDIAN E...E put into C

RMIN   E...E put into C

RMAX   E...E put into C

RSSQ   E...E put into C

TO FIND PROBABLITIES

PDF for values in E [put into E] calculates probabilities for the specified values of a discrete distribution and calculates the probability density function for a continuous distribution.

CDF for values in E...E [put into E...E] gives the cumulative distribution. For any value X, CDF X gives the probability that a random variable with the specified distribution has a value less than or equal to X.

INVCDF for values in E [put into E] gives the inverse of the CDF.

Each of these commands apply the following distributions (as well as some others). If no subcommand is used, the default distribution is the standard normal.

BINOMIAL	$n = K, p = K$
POISSION	$\mu = K$ (note that for the Poisson distribution, $\mu = \lambda$)
INTEGER	$a = K, b = K$
DISCRETE values in C, probabilities in C	
NORMAL	$\mu = K, \sigma = K$
UNIFORM	$a = Km\ b = K$
T	$d.f. = K$
F	$d.f$ numerator $= K, d.f.$ denominator $= K$
CHISQUARE	$d.f. = K$

WINDOWS menu selection: **➤Calc➤Probability Distribution➤Select distribution**

In the dialog box, select **Probability for PDF; Cumulative probability for CDF; Inverse cumulative for INV;** enter the required information such as **E, n, p,** or **μ, d.f.** and so forth.

GRAPHING COMMANDS

Character Graphics Commands

Note: In some versions of MINITAB, you must use the command **GSTD** before you use the following graphics commands.

PLOT C versus C prints a scatter plot with the first column on the vertical axis and the second on the horizontal axis. The following subcommands can be used with **PLOT.**

TITLE = 'text' gives a title above the graph.

FOOTNOTE = 'text' places a line of text below the graph.

XLABEL = 'text' labels the x-axis.

YLABEL = 'text' labels the y-axis.

SYMBOL = 'symbol' selects the symbol for the points on the graph. The default is *.

XINCREMENT = K is distance between tick marks on x-axis.

XSTART = K [end = k' specifies the first tick mark and optionally the last one.

YINCREMENT = K is distance between tick marks on y-axis.

YSTART = K [end = K] specifies the first tick mark and optionally the last one.

WINDOWS menu selection: **➤Graph➤Character Graphs➤Scatter Plot**

Titles, labels, and footnotes are in the **Annotate...** option.

Increment and start are in the **Scale** option.

Professional Graphics

Note: In some versions of MINITAB, you must use the command **GPRO** before you use the following graphics commands.

Plot C * C prints a scatter plot with the first column on the vertical axis and the second on the horizontal axis. Note that the columns must be separated by an asterisk *.

    Connect connects the points with a line

Other subcommands may be used to title the graph and set the tick marks on the axes. See your MINITAB software manual for details.

    WINDOWS menu selection: ➤**Graph**➤**Plot**

      Use the dialog boxes to title the graph, label the axes, set the tick marks, and so forth.

      See your MINITAB software manual for details.

CONTROL CHARTS

Character Graphics Commands

Note: In some versions of Minitab, you must use the command **GSTD** before you use the following graphics commands.

CHART C...C produces a control chart under the assumption that the data come from a normal distribution with mean and standard deviation specified by the subcommands.

MU = K   gives the mean of the normal distribution.

SIGMA = K gives the standard deviation.

    WINDOWS menu selection: none for character graphics. Use the commands in the session window.

Professional Graphics

Note: This may be the default mode for versions of MINITAB supporting professional graphics. If necessary, use the command **GPRO** before using the following commands.

CHART C...C produces a control chart under the assumption that the data come from a normal distribution with mean and standard deviation specified by the subcommands

MU K gives the mean of the normal distribution.

SIGMA K gives the standard deviation..

    WINDOWS menu selection: ➤**Stat**➤**Control Chart**➤**Individual**

      Enter choices for MU and Sigma in the dialog box.

TO GENERATE CONFIDENCE INTERVALS

ZINTERVAL [K% confidence] $\sigma = K$ **on C...C** generates a confidence interval for μ using the normal distribution. You must enter a value for σ, either actual or estimated. A separate interval is given for data in each column. If K is not specified, a 95% confidence interval will be given.

    WINDOWS menu selection: ➤**Stat**➤**Basic Statistics**➤**1-sample z**

      In the dialog box select confidence interval and enter the confidence level.

TINTERVAL [K% confidence] for C…C generates a confidence interval for μ using the Student's t distribution. It automatically computes stdev s from the data as well as the number of degrees of freedom. If K is not specified, a 95% confidence interval is given.

TO TEST A SINGLE MEAN

ZTEST [μ = K] σ = K, for C…C performs a z-test on the data in each column. If you do not specify μ, it is assumed to be 0. You need to supply a value for σ (either actual, or estimated by the sample standard deviation s of a column in the case of large samples). If the ALTERNATIVE subcommand is not used, a two-tailed test is conducted.

> WINDOWS menu selection: **≻Stat≻Basic Statistics≻1 sample z**

> In dialog box select alternate hypothesis, specify the mean for Hü, specify the standard deviation.

TTEST [μ = K] on C…C performs a separate t-test on the data of each column. If you do not specify μ, it is assumed to be 0. The computer evaluates s, the sample standard deviation for each column, and uses the computed s value to conduct the test. If the ALTERNATIVE subcommand is not used, a two-tailed test is conducted.

> WINDOWS menu selection: **≻Stat≻Basic Statistics≻1 sample t**

> In dialog box select alternate hypothesis, specify the mean for H_0.

ALTERNATIVE = K is the subcommand required to conduct a one-tailed test. If K = −1, then a left-tailed test is done. If K = 1, then a right-tailed test is done.

TO TEST A DIFFERENCE OF MEANS (INDEPENDENT SAMPLES)

TWOSAMPLE [K% confidence] for C…C does a two (independent) sample t test and (optional confidence interval) for data in the two columns listed. The first data set is put into the first column, and the second data set into the second column. Unless the ALTERNATIVE subcommand is used, the alternate hypothesis is assumed to be $H_1: \mu_1 \neq \mu_2$. Samples are assumed to be independent.

ALTERNATIVE = K is the subcommand to change the alternate hypothesis to a left-tailed test with K = −1 or right-tailed test with K = 1.

POOLED is the subcommand to be used only when the two samples come from populations with equal standard deviations.

> WINDOWS menu selection: **≻Stat≻Basic Statistics≻2 sample t**

> In dialog box select alternate hypothesis, specify the mean for H_0, for small samples select equal variances.

TO PERFORM SIMPLE OR MULTIPLE REGRESSION

REGRESS C on K explanatory variables in C…C does regression with the first column containing the response variable, K explanatory variables in the remaining columns.

PREDICT E…E predicts the response variable for the given values of the explanatory variable(s).

> WINDOWS menu selection: **➤Stat➤Regression➤Regression**

>> Use the dialog box to list the response and explanatory (prediction) variables. Mark the residuals box. In the Options dialog box list the values of the explanatory variable(s) for which you wish to make a prediction. Select the P.I. confidence interval.

BRIEF K controls the amount of output for K = 1, 2, 3 with 3 giving the most output. This command is not available from a menu.

There are other subcommands for REGRESS. See the MINITAB reference manual for your release of MINITAB for a list of the subcommands and their descriptions.

TO FIND THE PEARSON PRODUCT MOMENT CORRELATION COEFFICIENT

CORRELATION for C…C calculates the correlation coefficient for all pairs of columns.

> WINDOWS menu selection: **➤Stat➤Basic Statistics➤Correlation**

TO GRAPH THE SCATTER PLOT FOR SIMPLE REGRESSION

With **GSTD** use the **PLOT C vs C** command.

> WINDOWS menu selection (MINITAB SE for Windows or MINITAB 10 for Windows):
> **➤Stat➤Regression➤Fitted Line Plot**

TO PERFORM CHI SQUARE TESTS

CHISQUARE test on table stored in C…C produces a contingency table and computes the sample chi-square value

> WINDOWS menu selection: **➤Stat➤Tables➤Chisquare Test**

>> In the dialog box specify the columns which contain the chi-square table.

PART III

ComputerStat Guide

for

Understanding Basic Statistics, Third Edition

CHAPTER 1 GETTING STARTED

ABOUT COMPUTERSTAT

ComputerStat is a collection of computer programs designed for beginning statistics students and class demonstrations. The programs, together with this *Guide* constitute an educational tool rather than a commercial or research tool. They are designed to demonstrate statistical concepts as well as to process data.

There are more than 50 class demonstrations available in *ComputerStat*. Most of these class demonstrations use real data from designated references. A list of all the data files included in the class demonstrations can be found in the Appendix of this *Guide*.

The Lab Activities in this *Guide* support active classroom discussion. The *Guide* and programs are written so that very little classroom time need be devoted to learning how to use *ComputerStat*. Students also can work in a computer lab and gain useful insights from working with *ComputerStat* and the Lab Activities.

ComputerStat is not a tutorial package. Rather, it gives students an opportunity to further explore concepts they mastered in class without the necessity of doing a lot of calculations by hand.

Some advantages of *ComputerStat* are:

➤ *ComputerStat*'s easy to use class demonstrations can be selected from a menu containing real data from referenced sources. Students also may elect to enter their own data directly, as instructed by the program.

➤ The programs and Lab Activities are directly related to the corresponding units of *Understanding Basic Statistics* third edition.

➤ The programs together with the Lab Activities help students explore what happens when various conditions of a problem are modified.

➤ *ComputerStat* is inexpensive, does not require extensive equipment, and does not require extensive time to learn to use.

ComputerStat is housed on the new HM StatPass CD-ROM, which comes packaged for free with the third edition textbook. Please see the following system requirements and installation instructions for information on how to access *ComputerStat*.

Minimum System Requirements for HM StatPass CD-ROM

Windows	Macintosh
Pentium II	PowerPC
32MB AVAILABLE RAM	32 MB AVAILABLE RAM
4x speed CD-ROM drive	4x speed CD-ROM drive
800x600 screen resolution, hi-color mode	800x600 screen resolution, hi-color mode
Windows 95, 98, ME, NT, 2000, XP	OS 8.6, 9.x, X (in classic mode only)
Windows compatible sound card	Sound card
15 MB of hard disk space is required for a full install.	15 MB of hard disk space is required for a full install.
Internet Explorer 5	Internet Explorer 5
Netscape 4.6x	Netscape 4.6x

Installation Instructions

You have a choice of installing this program two ways. The default installation is a compact install, and the program runs from the CD-ROM. This requires that the CD-ROM is in the drive while using this program.

The second option is to install the program fully. The program requires approximately 15 MG of hard drive space for the full install. *Please note that performing a full install may take as much as 25 minutes.* Once the program is fully installed, you will not need to have the CD-ROM in the drive while using the program.

Installing HM StatPass for Windows

1. Insert the CD-ROM into the CD-ROM drive.
2. Double-click the "My Computer" icon located on your desktop, and then double-click on the CD-ROM drive icon, which should be labeled "StatPass."
3. Double-click the icon labeled "Setup.exe" to run the installer application.
4. Follow the instructions on the screen.
5. When installation is complete select "Close."
6. Launch the program from the Programs menu by going to Start/Programs/StatPass, and then selecting "Welcome.html."

Installing HM StatPass for Macintosh

1. Insert the CD-ROM into the CD-ROM drive.
2. Double-click the StatPass CD-ROM icon located on the desktop.
3. Double-click the StatPass Installer, select the location to install and follow the directions.
4. When installation is complete select "Quit."
5. Launch the program by navigating to the location you chose to install the program in and select the start.html file to launch using Internet Explorer or the startn.html file to launch using Netscape Navigator.

Running ComputerStat

ComputerStat is a stand-alone program included within the HM StatPass CD-ROM. You can access it from the Start Menu on your Windows machine, or from the folder where StatPass is installed on your Macintosh. Instructions for how to access the program can also be found in the StatPass program. Please note that when running the simulation of tossing two dice in the ComputerStat program on a Macintosh computer, if N is set to 1000000 (the maximum value), several of the values for Number of Occurrences are negative. This is a program bug.

As always, feel free to contact Houghton Mifflin Technical Support with any questions or concerns at 1-800-732-3223 or support@hmco.com.

USING COMPUTERSTAT

After starting *ComputerStat*, you will see opening title screens. Follow the instructions and click on the NEXT button until the main menu appears.

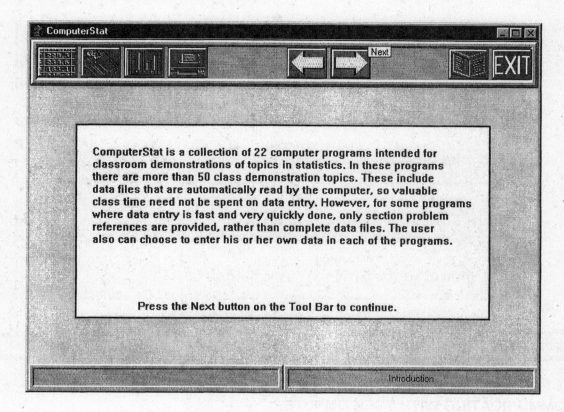

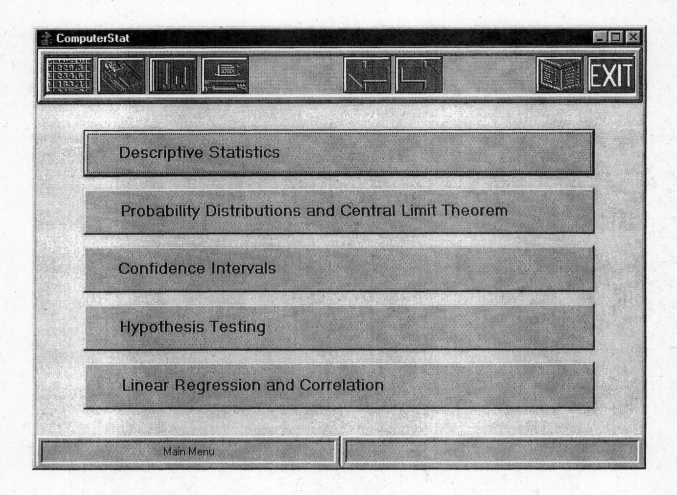

COMMUNICATING WITH THE COMPUTER

As you use *ComputerStat*, you will find that in most cases the computer wants you to respond with a number. Every time you wish to type a number, use the key with the desired digit. The period serves as a decimal point.

Using the mouse: Generally you will left-click the mouse to activate a menu item or to activate a response box.

NOTATION ON THE COMPUTER

Some mathematical symbols are expressed in a slightly different format on a computer.

Mathematical Notation	Computer Notation
$\leq$ (less than or equal to)	<=
$\geq$ (greater than or equal to)	>=
$\neq$ (not equal)	<>

Occasionally numbers will be expressed in scientific notation, such as 3.012758E-02. This notation tells us we are to multiply the number in front of the E by the power of 10 listed after the E.

$3.012758E\text{-}02 = 0.03012758$ The E-02 tells us to multiply 3.01278 by 10^{-2}.
This has the effect of moving the decimal point two places to the left.

$5.738613E\text{+}03 = 5738.613$ The E+03 part tells us to multiply 5.738613 by 10^3.
This has the effect of moving the decimal point three places to the right.

In short, if the value following E is

negative, move the decimal that many places to the *left*.

positive, move the decimal that many places to the *right*.

RANDOM SAMPLES
(SECTION 1.2 OF *UNDERSTANDING BASIC STATISTICS*)

Main menu selection: Descriptive Statistics
Sub-menu selection: Random Samples

I. Description of the program

In this program a random number generator is used to do the following activities:

(a) Select a random sample of size M from a population of size N and list the chosen sample items. You may sample *with* or *without* replacement.

(b) Simulate the experiment of tossing one die up to 5000 times.

(c) Simulate the experiment of tossing two dice up to 5000 times.

Input: First you select the option you wish to run.

Option 1:

Select a random sample of size M from a population of size N.

Population size N; N between 1 and 1,000,000

Sample size M; M between 1 and 500 for sampling with replacement; M between 1 and the minimum of N or 500 for sampling without replacement.

Choices available: sampling with or without replacement

Option 2:

Simulate the experiment of tossing one die.

Number of times to toss die, N; N between 1 and 5000

Option 3:

Simulate the experiment of tossing two dice.

Number of times to toss dice, N; N between 1 and 5000

Finally, you have a choice of repeating any of the three options, returning to the sub-menu, or returning to the main menu.

Output: For the three options,

Option 1:

The numbers corresponding to the data elements from the population that are to be included in the sample.

Option 2:

Number of occurrences of each of the 6 possible outcomes of tossing one die, and the relative frequencies of the outcomes.

Option 3:

Number of occurrences of each of the 36 possible outcomes of tossing two dice, and the corresponding relative frequencies.

Number of occurrences of each of the 12 possible sums on the two dice, with corresponding relative frequencies.

II. Sample Run

Suppose there are 75 students enrolled in a large section of statistics. Draw a random sample of 15 students without using any student in the sample more than once.

First assign each of the 75 students a number from 1 to 75. The population size is 75. The sample size is 15. Instruct the computer to sample without replacement, since you want each of the students in the sample to be different. The computer will give you a list of numbers. These are the numbers assigned to the students to be included in your sample. From the student numbers, you can determine which students are in your sample.

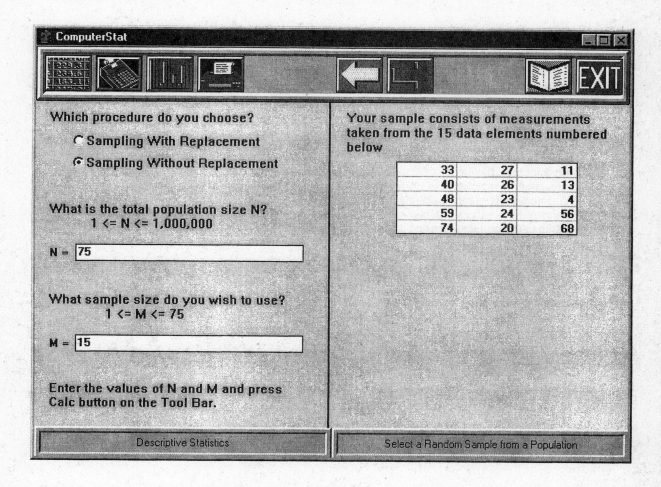

LAB ACTIVITIES FOR RANDOM SAMPLES

1. Out of a population of 8000 eligible county residents, select a random sample of 50 for prospective jury duty. Should you sample with or without replacement? Explain.

2. Simulating experiments in which outcomes are equally likely is another important use of random numbers.

   We can simulate dealing bridge hands by numbering the cards in a bridge deck from 1 to 52. Then we draw a random sample of 13 numbers without replacement from the population of 52 numbers. A bridge deck has 4 suits: hearts, diamonds, clubs, and spades. Each suit contains 13 cards: those numbered 2 through 10, a jack, a queen, a king, and an ace. Decide how to assign the numbers 1 through 52 to the cards in the deck. Use the *ComputerStat* program Random Samples to get the numbers of the 13 cards in one hand. Translate the numbers to specific cards and tell what cards are in the hand. For a second game the cards would be collected and reshuffled. Rerun the program option and determine the hand you might get in a second game.

3. We can also simulate the experiment of tossing a fair coin. The possible outcomes resulting from tossing a coin are heads or tails. Assign the outcome heads the number 2 and the outcome tails the number 1. Use the *ComputerStat* program Random Samples to simulate the act of tossing a coin 10 times. Note that you will use a population size of 2, since that is the number of outcomes. You need to sample with replacement. Select a sample of size 10 since you will be tossing the coin 10 times. Record the numbers of the outcomes and then record the outcomes themselves. Repeat the process for 30 trials.

4. When we toss a die, there are six possible outcomes: the number of dots appearing on the top will be 1, 2, 3, 4, 5 or 6. If the die is fair, each of the outcomes is equally likely to occur. However, if you toss the die 6 times, you will not necessarily get all six outcomes. It is quite possible to get two or three of the outcomes several times, and not get other outcomes at all. Run Random Samples and select Option 2, which simulates the experiment of tossing one die. Record the outcomes of tossing the die N = 6 times. Are the outcomes fairly evenly distributed? When you study probability (Chapter 5 of *Understanding Basic Statistics*) you will gain an understanding of why the outcomes of the experiments can vary. We will return to this program then and use the dice experiments to gain insights into probability.

CHAPTER 2 ORGANIZING DATA

FREQUENCY DISTRIBUTIONS
(SECTION 2.2 OF *UNDERSTANDING BASIC STATISTICS*)

Main menu selection: Descriptive Statistics
Sub-menu selection: Frequency Distributions and Grouped Data

I. Description of the program

This program produces a frequency distribution, relative frequency distribution, and cumulative frequency distribution with up to 10 classes. In addition it shows the corresponding histogram and the corresponding ogive. The frequency table shows class boundaries, class frequencies, relative frequencies, class midpoints, and cumulative frequencies. In addition, there is an option for computing the mean and standard deviation from both the raw data and the grouped data.

Input:

Option 1: Select a class demonstration.

Use an existing class demonstration data file that is included in the data menu of *ComputerStat*. The data values included in the demonstrations are displayed in the Appendix of this guide.

Enter number of classes, N; N between 1 and 10.

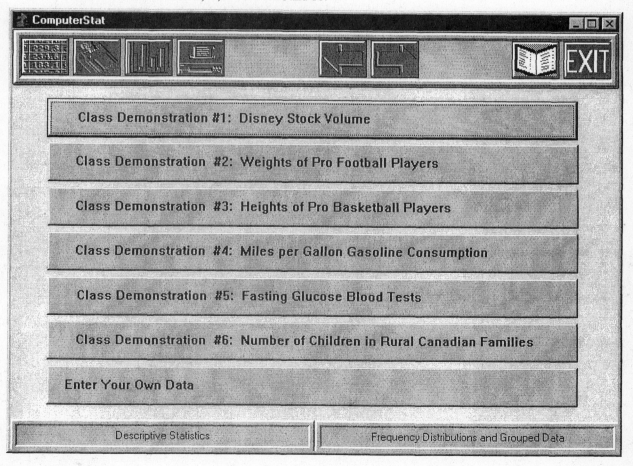

Option 2: Enter your own data.

Number of digits to the right of the decimal point; this value can be a number from 0 through 4. Scan all your data and find the value with the most places after the decimal point. This is the number you will need to tell the computer. If you have more than 4 digits after the decimal point, you will either need to round your data or rescale (multiply by a power of 10) it before entering it into the program.

Number of pieces of data, M; M between 2 and 200

Number of classes, N; N between 1 and 10

Ith data value X(I) if you enter your own data; If you make an error when entering the data, jot down the value of I where the error occurred.

Data correction option; If you know the Ith position in which the data error occurred, you can correct it with this option.

Categorization of data type, sample or population; This input in necessary in the options to compute the mean and standard deviation.

Program Options:

(1) Rerun program with same data and new classes.
(2) Rerun program with new data.
(3) Graph histogram.
(4) Graph ogive.
(5) Compute mean and standard deviation of data.

Output:

List of data values
Smallest data value
Largest data value
Range
Class width

Frequency table: included are class boundaries, class frequencies, relative frequencies, class midpoints,
                    and cumulative frequencies
Histogram
Ogive
Mean and standard deviation computed from raw data
Mean and standard deviation estimated from the data as it appears in the frequency table.

II. Sample Run

Throughout the day from 8 A.M. to 11 P.M., Tiffany counted the number of ads occurring every hour on one commercial TV station. The 15 data values are

$$10 \quad 12 \quad 8 \quad 7 \quad 15 \quad 6 \quad 5 \quad 8$$
$$8 \quad 10 \quad 11 \quad 13 \quad 15 \quad 8 \quad 9$$

Use the *ComputerStat* program Frequency Distribution and Grouped Data to find the smallest and largest data values and the range. Using 4 classes, find the class width, make a frequency table, and graph the histogram.

First enter the data. Select Enter Your Own Data from the data menu.

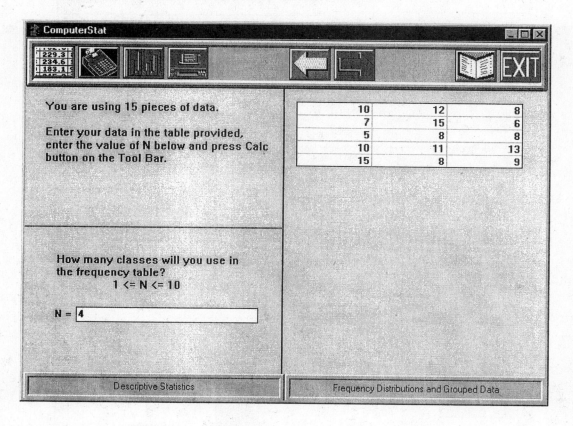

Click Calculate (the calculator).

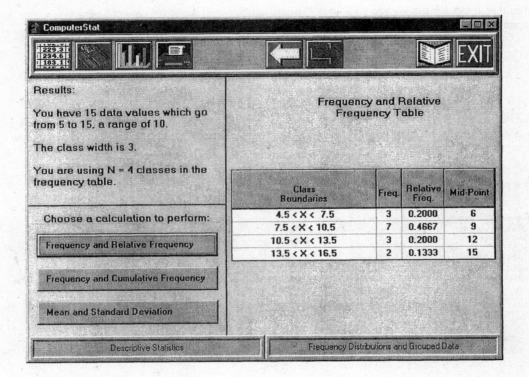

Click the Graph button for graphing options.

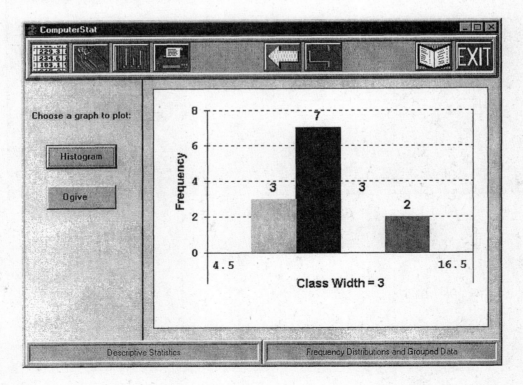

LAB ACTIVITIES FOR FREQUENCY DISTRIBUTIONS

1. The following data represent the number of shares of Disney stock (in hundreds of shares) sold for a random sample of 60 trading days in 1993 and 1994.

Source: Dow-Jones Information Retrevial Service

12584.00	9441.00	18960.00	21480.00	10766.00
13059.00	8589.00	4965.00	4803.00	7240.00
10906.00	8561.00	6389.00	14372.00	18149.00
6309.00	13051.00	12754.00	10860.00	9574.00
19110.00	29585.00	21122.00	14522.00	17330.00
18119.00	10902.00	29158.00	16065.00	10376.00
10999.00	17950.00	15418.00	12618.00	16561.00
8022.00	9567.00	9045.00	8172.00	13708.00
11259.00	10518.00	9301.00	5197.00	11259.00
10518.00	9301.00	5197.00	6758.00	7304.00
7628.00	14265.00	13054.00	15336.00	14682.00
27804.00	16022.00	24009.00	32613.00	19111.00

Use the *ComputerStat* Program Frequency Distributions And Grouped Data. Select Class Demonstration #1 Disney Stock Volume. Use the program to make a frequency table showing class boundaries, frequencies, relative frequencies, midpoints, and cumulative frequencies for 5 classes. Repeat the process for 9 classes. Is the data distribution skewed or symmetric? Is the distribution shape more pronounced with 5 classes or with 9 classes? Look at the ogive for 9 classes. Approximate the volume level traded on 75% of the days.

2. Explore the other class demonstration data files in the options menu for Frequency Distributions.

Class Demonstration #2: Weights of Pro Football Players
Class Demonstration #3: Heights of Pro Basketball Players
Class Demonstration #4: Miles per Gallon Gasoline Consumption
Class Demonstration #5: Fasting Gluose Blood Tests
Class Demonstration #6: Number of Children in Rural Canada Families

In each case:

(a) Make histograms with 5, 8, and 10 classes.
(b) Categorize the shape of the distribution as: uniform, symmetric, skewed, or bimodal.

3. (a) Consider the data

$$
\begin{array}{ccccc}
1 & 3 & 7 & 8 & 10 \\
6 & 5 & 4 & 2 & 1 \\
9 & 3 & 4 & 5 & 2
\end{array}
$$

Use the *ComputerStat* program Frequency Distributions and Grouped Data to make a frequency table and histogram with N = 3 classes. Jot down the results so you can compare them to part (b).

(b) Now add 20 to each data value of part (a). The results are

$$
\begin{array}{ccccc}
21 & 23 & 27 & 28 & 30 \\
26 & 25 & 24 & 22 & 21 \\
29 & 23 & 24 & 25 & 22
\end{array}
$$

Make a frequency table with 3 classes. Compare the class boundaries, midpoints, frequencies, relative frequencies, and cumulative frequencies with those obtained in part (a). Are each of the boundaries and midpoints 20 more than those of part (a)? How do the frequencies, relative frequencies, and cumulative frequencies compare?

(c) Use your discoveries from part (b) to predict the frequency table and histogram with 3 classes for the data values below.

$$
\begin{array}{ccccc}
1001 & 1003 & 1007 & 1008 & 1010 \\
1006 & 1005 & 1004 & 1002 & 1001 \\
1009 & 1003 & 1004 & 1005 & 1002
\end{array}
$$

Would it be safe to say that we simply shift the histogram of part (a) 1000 units to the right?

(d) What if we multiply each of the values of part (a) by 10? Will we effectively multiply the entries for class boundaries by 10? To explore this relation, use Frequency Distributions to generate the histogram and frequency table with 3 classes for the data.

$$
\begin{array}{ccccc}
10 & 30 & 70 & 80 & 100 \\
60 & 50 & 40 & 20 & 10 \\
90 & 30 & 40 & 50 & 20
\end{array}
$$

Compare the frequency table of these data to the one from part (a). You will see that there does not seem to be an exact correspondence. To see why, look at the class width and compare it to the class width of part (a). The class width is always increased to the next integer value no matter how large the integer data values are. Consequently, the class width for the data in part (d) was increased to 31 instead of to 40.

4. Histograms are not effective displays for some data. Consider the data

1	2	3	6	5	7
9	8	4	12	11	15
14	13	6	2	1	206

Use Frequency Distributions to create frequency tables and histograms with 2 classes. Then change to 3 classes, on up to 10 classes. Notice that all the histograms lump the first 17 data values into the first class, and the one data value 206 in the last class. What feature of the data causes this phenomenon? Recall that

Class width = (largest value − smallest value)/(number of classes) *increased to the next integer*

How many classes would you need before you began to see the first 17 data values distributed among several classes? What would happen if you simply did not include the extreme value 206 in your histogram?

Chapter 3 AVERAGES AND VARIATION

AVERAGES, VARIATION, BOX-AND-WHISKER PLOTS
(SECTIONS 3.1–3.3 OF *UNDERSTANDING BASIC STATISTICS*)

Main Menu selection: Descriptive Statistics
Sub-menu selection: Averages, Variation, Box-and-Whisker Plots

I. Description of the program

This program computes averages and measures of variation for raw data. The averages include median, mode, and mean. Measures of variation include the range, standard deviation and variance computed for data forming a sample or population. The program also produces the five-number summary used to make a box-and-whisker plot and displays the graph of a box-and-whisker plot.

Input:

Option 1: Select a class demonstration.

Use an existing class demonstration data file included in the *ComputerStat* data menu. The data values included in the demonstrations are displayed in the Appendix of this guide.

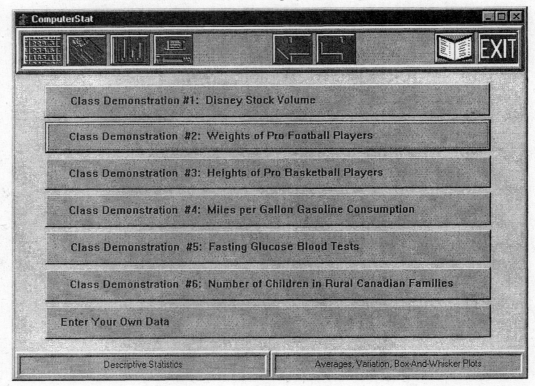

Option 2: Enter your own data.

Number of pieces of data N; N between 2 and 100

Data type; Forming a sample or forming a population

Ith data value, X(I); Again, note the Ith position of any data entry error for later correction.

Data correction option; If you know the Ith position in which the data error occurred, you can correct it with this option.

Output:

List of data ordered from smallest to largest

Mode

Mean

Range

Variance; Sample or population

Standard deviation and CV; Sample or population

Five-number summary for box-and-whisker plot

low value

first quartile value

median

third quartile value

high value

Box-and-whisker plot

II. Sample Run

Consider the weights of professional football players. The weights of a random sample of 50 linebackers from professional teams are included in Class Demonstration #2. Use the program to generate a statistical summary of the weights and to draw a box-and-whisker plot.

Run the program, and select Class Demonstration #2: Weights of Pro Football Players. You will see a data screen listing the data. When you click on the Calculate button (calculator key) this screen appears. Click on Mode, Mean, Median to get the right side.

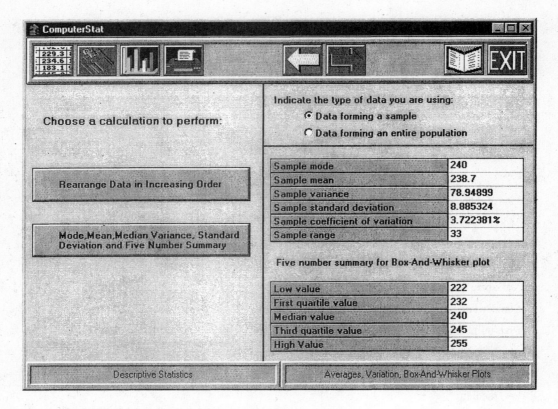

Click on the Graph button to get a box-and-whisker plot.

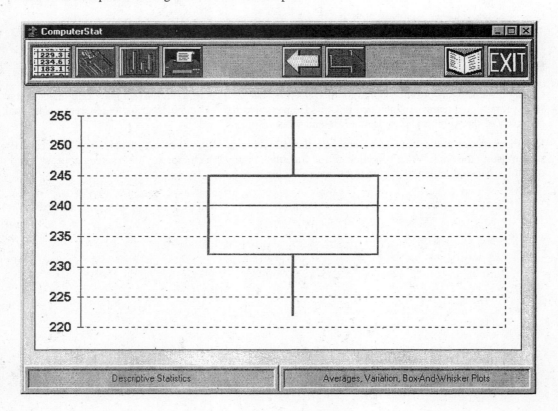

LAB ACTIVITIES FOR AVERAGES, VARIATION, BOX-AND-WHISKER PLOTS

1. Select one of the class demonstrations available in the options menu:

(1) Class Demonstration #1: Disney Stock Volume

(2) Class Demonstration #2: Weights of Pro Football Players

(3) Class Demonstration #3: Heights of Pro Basketball Players

(4) Class Demonstration #4: Miles per Gallon Gasoline Consumption

(5) Class Demonstration #5: Fasting Gluose Blood Tests

(6) Class Demonstration #6: Number of Children in Rural Canadian Familes

Use the program Averages, Variation, Box-and-Whisker Plots to generate a statistical summary of the data. Then, use the program Frequency Distributions and Grouped Data with the same class demonstration to generate a histogram and ogive with 7 classes.

Use the information from both programs to answer the following questions:

(a) Is the distribution skewed or symmetric? How is this shown in both the histogram and the box-and-whisker plot?

(b) Look at the box-and-whisker plot. Are the data more spread out above the median or below the median?

(c) Look at the histogram and estimate the location of the mean on the horizontal axis. Are the data more spread out above the mean or below the mean?

(d) Do there seem to be any data values that are unusually high or unusually low? If so, how do these show up on a histogram or on a box-and-whisker plot?

(e) Pretend you are writing a brief article for a newspaper. Describe the information about the data in the class demonstration you selected in non-technical terms. Be sure to make some comments about the "average" of the data measurements and some comments about the spread of the data.

2. (a) Consider the test scores of 30 students in a political science class:

85	73	43	86	73	59	73	84	62	100
75	87	70	84	97	62	76	89	90	83
70	65	77	90	84	80	68	91	67	79

For this population of test scores, find the mode, median, mean, range, variance, standard deviation, *CV*, and the five-number summary and make a box-and-whisker plot. Be sure to record all of these values so you can compare them to the results of part (b).

(b) Suppose Greg was in the political science class of part (a). Suppose he missed a number of classes because of illness, but took the exam anyway and made a score of 30 instead of 85 as listed as the first entry of the data in part (a). Again, use the program to find the mode, median, mean, range, variance, standard deviation, *CV*, and the five-number summary and make a box-and-whisker plot using the new data set. Compare these results to the corresponding results of part (a). Which average was most affected: mode, median, or mean? What about the range, standard deviation, and coefficient of variation? How do the box-and-whisker plots compare?

(c) Write a brief essay in which you use the results of parts (a) and (b) to predict how an extreme data value affects a data distribution. What do you predict for the results if Greg's test score had been 80 instead of 30 or 85?

3. In this problem we will explore the effects of changing data values by multiplying each data value by a constant, or by adding the same constant to each data value.

(a) Consider the data

1	8	3	5	7
2	10	9	4	6
3	5	2	9	1

Use the computer program to find the mode (if it exists), mean, sample standard deviation, range, and median and box-and-whisker plot. Make a note of these values, since you will compare them to those obtained in parts (b) and (c).

(b) Now multiply each data value of part (a) by 10 to obtain the data

10	80	30	50	70
20	100	90	40	60
30	50	20	90	10

Again use the computer program to find the mode (if it exists), mean, sample standard deviation, range, and median and box-and-whisker plot. Compare these results to the corresponding ones of part (a). Which values changed? Did those that changed change by a factor of 10? Did the range or standard deviation change? Referring to the formulas for these measures (see Section 3.2 of *Understanding Basic Statistics*) can you explain why these values behaved the way they did? Will these results generalize to the situation of multiplying each data entry by 12 instead of by 10? What about multiplying each by 0.5? Predict the corresponding values that would occur if we multiplied the data set of part (a) by 1000.

(c) Now suppose we add 30 to each data value part (a).

32	38	33	35	37
32	40	39	34	36
33	35	32	39	31

Again use the computer program to find the mode (if it exists), mean, sample standard deviation, range, and median and box-and-whisker plot. Compare these results to the corresponding ones of part (a). Which values changed? Of those that are different, did each change by being 30 more than the corresponding value of part (a)? Again look at the formulas for range and standard deviation. Can you predict the observed behavior from the formulas? Can you generalize these results? What is we added 50 to each data value of part (a). Predict the values for the mode, mean, sample standard deviation, range, and median.

CHAPTER 4 REGRESSION AND CORRELATION

LINEAR REGRESSION AND CORRELATION
(SECTIONS 4.1–4.3 OF *UNDERSTANDING BASIC STATISTICS*)

Main Menu selection: Linear Regression and Correlation
Sub-menu selection: Linear Regression and Testing the Correlation Coefficient Rho
Main Menu selection: Hypothesis Testing
Sub-menu selection: Linear Regression and Testing the Correlation Coefficient Rho

I. Description of the program

The program Linear Regression and Correlation performs a number of function related to linear regression and correlation. The user inputs a random sample of ordered pairs of X and Y values. Then the computer returns the sample means, sample standard deviations of X and Y values; largest and smallest X value; standard error of estimate; equation of least-squares line; the correlation coefficient and the coefficient of determination.

Next the user has several options.

1. Predict Y values from X values, and when the X value is in the proper range, compute C% confidence intervals for Y values(a topic in Section 12.4 of *Understanding Basic Statistics*).

2. Test the correlation coefficient, a topic in Section 12.5 of *Understanding Basic Statistics*. If we let RHO represent the population correlation coefficient, then the null hypothesis is H0: RHO = 0. The alternate hypothesis may be one of the following.

$$\text{H1: RHO} < 0 \text{ (left-tailed test)}$$
$$\text{H1: RHO} > 0 \text{ (right-tailed test)}$$
$$\text{H1: RHO} <> 0 \text{ (two-tailed test)}$$

3. Graph data points (X, Y) to create a scatter diagram. Then the user can see five points on the least-squares line.

Input:

Option 1: Select a class demonstration:

(1) Class Demonstration #1: List Price vs. Best Price For A New GMC Pickup Truck

(2) Class Demonstration #2: Cricket Chirps vs. Temperature

(3) Class Demonstration #3: Diameter of Sand Granules vs. Slope on a Naturally Occurring Ocean Beach

(4) Class Demonstration #4: National Unemployment Rate Male vs. Female

Option 2:

Value of X for which you want to predict Y

Confidence level C for confidence interval of Y values

Alternate hypothesis for testing RHO (Section 12.5 of *Understanding Basic Statistics*)

Level of significance for testing RHO (Section 12.5 of *Understanding Basic Statistics*)

Output:

Information summary table including the least squares line, S_e, R, R^2

For a given X value, the predicted Y value and a C% confidence interval

Test summary for testing RHO (Section 12.5 of *Understanding Basic Statistics*)

Scatter diagram and least-squares line

II. Sample Run

Begin the program Linear Regression and Testing the Correlation Coefficient Rho. Click on the Data button. Select Class Demonstration #2) Cricket Chirps vs. Temperature. Use the program to explore the relationship between cricket chirp frequency and temperature.

After you select the class demonstration, the data is shown followed by the information summary.

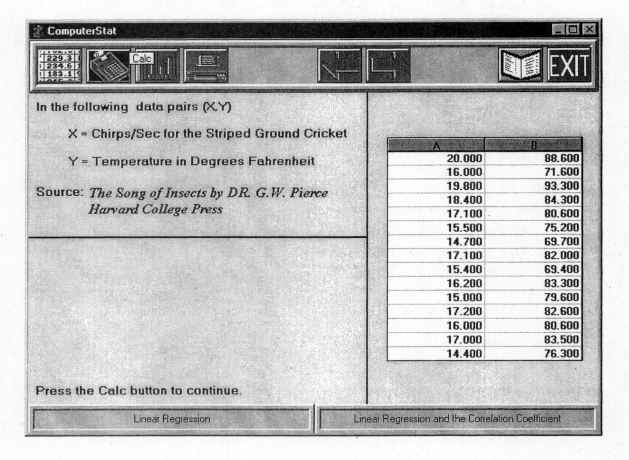

Click on the calculate button.

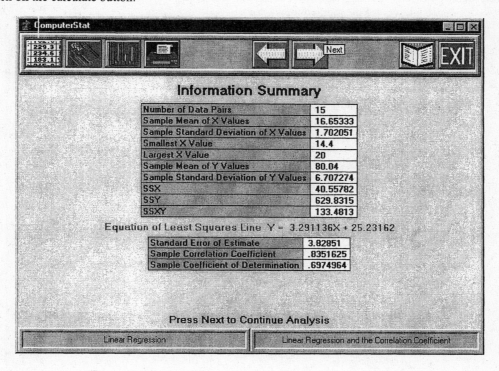

Clicking on the Next button gives you a choice of graphing the scatter diagram and least-squares line, predicting Y from an X value, or testing the correlation coefficient R. Click on the Graph button at the top of the window.

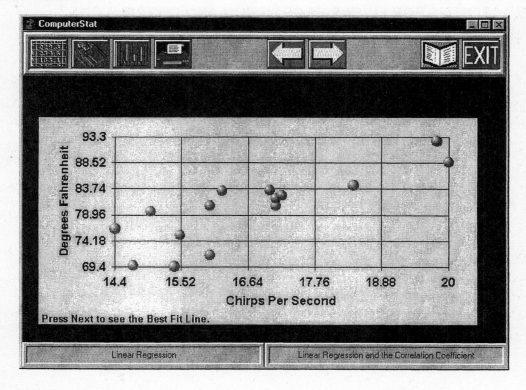

Click on the "previous" button (left arrow) twice, and then select the Prediction option. For X = 18 and a 90% confidence interval for the prediction, we see

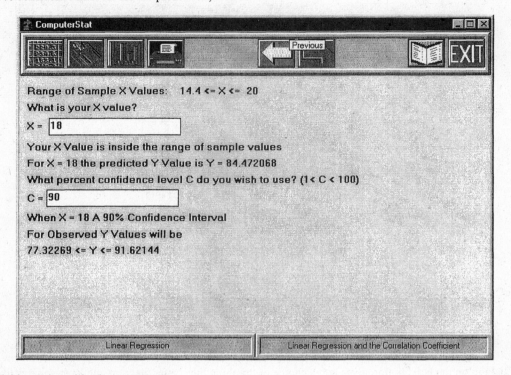

Clicking Print will print your results.

LAB ACTIVITIES FOR LINEAR REGRESSION AND CORRELATION

1. Select Class Demonstration #1: List Price vs. Best Price for a New GMC Pickup Truck.

> In the Following Data Pairs (X, Y),
> X = List Price (in $1000) for a GMC Pickup Truck
> Y = Best Price (in $1000) for a GMC Pickup Truck

(12.400, 11.200)	(14.300, 12.500)	(14.500, 12.700)
(14.900, 13.100)	(16.100, 14.100)	(16.900, 14.800)
(16.500, 14.400)	(15.400, 13.400)	(17.000, 14.900)
(17.900, 15.600)	(18.800, 16.400)	(20.300, 17.700)
(22.400, 19.600)	(19.400, 16.900)	(15.500, 14.000)
(16.700, 14.600)	(17.300, 15.100)	(18.400, 16.100)
(19.200, 16.800)	(17.400, 15.200)	(19.500, 17.000)
(19.700, 17.200)	(21.200, 18.600)	

(a) Look at the scatter diagram. Do you think linear regression is appropriate for this data?

(b) Look at the information summary sheet and write down the equation of the least-squares line, the value of the correlation coefficient, the value of the coefficient of determination, and the value of the standard error of estimate.

(c) If the list price of a pickup truck is $20,000, what is the predicted best price? (Note: Be sure to change 20,000 to 20 thousand, since the data have been entered in thousands.) Would you be surprised if the best price you could get were $19,500?

2. Merchandise loss due to shoplifting, damage, and other causes is called shrinkage. Shrinkage is a major concern to retailers. The managers of H.R. Merchandise think there is a relationship between shrinkage and number of clerks on duty. To explore this relationship, a random sample of 7 weeks was selected. During each week the staffing level of sale clerks was kept constant and the dollar value of the shrinkage was recorded. The results follow.

Number of clerks, X	10	12	11	15	9	13	8
Shrinkage, Y (in $hundreds)	19	15	20	9	25	12	31

(a) Select the option to enter your own data. Enter the data. Look at the scatter diagram. Does a linear regression model seem appropriate for this model?

(b) Look at the information summary sheet and write down the equation of the least-squares line, the value of the correlation coefficient, the value of the coefficient of determination, and the value of the standard error of estimate.

(c) Does it make sense to extrapolate beyond the data? Why or why not? What does the model predict for shrinkage if there is a staff of 40 clerks? What about for a staff of 1 clerk? Can you get confidence intervals for these predictions?

(d) What does the model predict for a staffing level of 12 clerks? Find a 90% confidence interval for shrinkage.

CHAPTER 5 ELEMENTARY PROBABILITY THEORY

SIMULATION OF PROBABILITY EXPERIMENTS
(SECTIONS 5.1 AND 5.2 OF *UNDERSTANDING BASIC STATISTICS*)

Main Menu selection: Descriptive Statistics
Sub-menu selection: Random Samples

There are three programs within this menu option:
Select a Random Sample from a Population
Simulate the Experiment of Tossing One Die
Simulate the Experiment of Tossing Two Dice

In Chapter 1, we showed how to use *ComputerStat* to select a random sample from a population.

I. Description of the program Simulate the Experiment of Tossing one Die

Input: The number of times you want to toss the die (from 1 to 1,000,000 times)

Output: The number of each of the outcomes 1 through 6, and the relative frequency of each outcome

II. Sample Run: Simulate tossing a die 400 times.

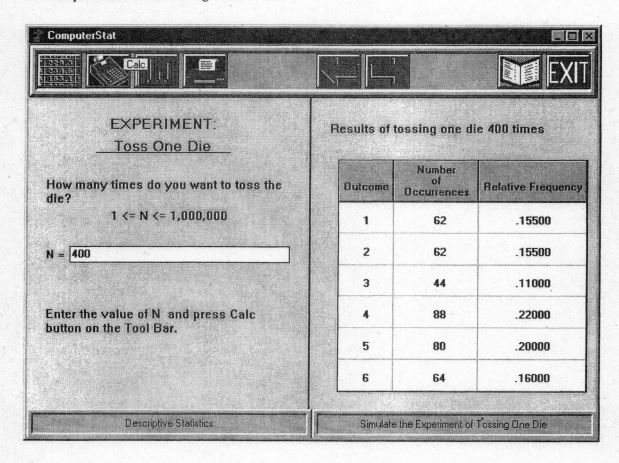

EXPERIMENT:
Toss One Die

How many times do you want to toss the die?

$1 <= N <= 1,000,000$

N = 400

Enter the value of N and press Calc button on the Tool Bar.

Results of tossing one die 400 times

Outcome	Number of Occurrences	Relative Frequency
1	62	.15500
2	62	.15500
3	44	.11000
4	88	.22000
5	80	.20000
6	64	.16000

Descriptive Statistics

Simulate the Experiment of Tossing One Die

LAB ACTIVITIES FOR PROBABILITY EXPERIMENTS

1. In the program Random Samples, select the option to simulate the experiment of tossing one die. Since the possible outcomes 1, 2, 3, 4, 5, 6 are equally likely, the theoretical probability for each outcome is 1/6 or approximately 0.16667.

   (a) Choose N = 6 tosses. Did you get an outcome of 2? How many times? Repeat the experiment again with N = 6. Did you get an outcome of 2 this time? Did it occur as many times as before? Repeat the experiment a third time. Should you expect the same results as you got before? Does the fact that the theoretical probability $P(\text{outcome is } 2) = 1/6$ guarantee that every time you toss a die six times you will get exactly one outcome of 2? Why or why not?

   (b) Repeat the experiment using number of tosses N = 50, 500, 1000, and 3000. Record the relative frequency for the outcome of 2. How do the relative frequencies compare to the theoretical results $P(\text{outcome is } 2) = 0.1667$? Does it appear that more trials lead to results that match the theory better?

   (c) When you toss a die, the outcomes are mutually exclusive, since it is impossible to show two distinct numbers on the top of one die. Therefore, theoretically, $P(\text{even outcome}) = P(2 \text{ or } 4 \text{ or } 6) = P(2) + P(4) + P(6) = 0.5$.

   Run the option to toss one die. Let the number of tosses be 5000. Using the number of occurrences of an outcome, compute the probability of obtaining a 2 or 4 or 6. In other words, add up the total occurrences of even numbers and divide by the total number of trials, 5000. How does this result compare to the theoretical result of 0.5? Do you get the same experimental result if you add the relative frequencies of obtaining a 2 or 4 or 6? Does the addition rule seem to work for mutually exclusive events?

2. In the program Random Samples, select the option to simulate the experiment of tossing two dice. Use N = 1000 tosses.

   (a) Use relative frequencies to estimate $P(2 \text{ on first die } and \text{ 5 on second})$. How does this result compare to the theoretically computed result using the multiplication rule for independent events?

   (b) Repeat part (a) for 5000 tosses. Do the experimental and theoretical results agree better?

   (c) Repeat parts (a) and (b) for $P(6 \text{ on first die } and \text{ odd on second})$.

3. In the program Random Samples, the option to simulate the experiment of tossing two dice also gives the relative frequencies of the possible sums. Let the number of tosses be N = 1000.

   (a) Use one of the screens showing the 36 possible outcomes possible when you toss two dice once to estimate the theoretical probabilities

   $$P(\text{sum of } 8), P(\text{sum of } 9), P(\text{sum of } 10), P(\text{sum of } 11), P(\text{sum of } 12)$$

   (b) Continue with the program until you see the screen showing the possible sums and their relative frequencies. How do these numbers compare to the results you got in part (a).

   (c) Since the events are mutually exclusive, we can compute

   $$P(\text{sum more than } 10) = P(\text{sum of } 11) + P(\text{sum of } 12)$$

   Using the results of part (a), estimate the theoretical probability of a sum more than 10.

   (d) Using the computer output, estimate the probability of getting a sum of more than 10 by adding the number of occurrences of sums more than 10 and dividing by the number of times the dice were tossed, 1000. How does this result compare to the estimated probability of part (c)?

   (e) Repeat parts (a) through (d) using $n = 5000$ trials.

4. Assume that the outcomes of having a boy or girl baby are equally likely. We can experimentally estimate the probability that a family with three children has all girls. In the program Random Samples, select the option of choosing a sample from a population. Assign the outcome 1 to boy and 2 to girl, so our "population" consists of the two numbers 1 and 2 and N = 2. Sample with replacement, since the outcomes boy and girl are independent. Choose a sample size = 3 to reflect that we are looking at families with three children.

(a) Record the outcomes in the sample. The list 1, 2, 2 translates to boy, girl, girl.

(b) Repeat the experiment for 29 more families of three children and record the outcomes of each of the families. Each time use population size N = 2, sample with replacement, and use sample size M = 3.

(c) Now you have a total of 30 groups of 3 children. How many of these groups contain all girls? What is the experimental estimate for the probability that a family of three children consists of there girls? In Chapter 6 of *Understanding Basic Statistics*, you will see how to compute the theoretical probability of having three girls out of three children, and we ask you to do this in the Lab Activities for Binomial Probabilities.

CHAPTER 6 THE BINOMIAL PROBABILITY DISTRIBUTION AND RELATED TOPICS

EXPECTED VALUE AND STANDARD DEVIATION OF DISCRETE PROBABILITY DISTRIBUTIONS
(SECTION 6.1 OF *UNDERSTANDING BASIC STATISTICS*)

Main Menu selection: Probability Distributions and Central Limit Theorem
Sub-menu selection: Expected Value and Standard Deviation of Discrete Probability Distributions

I. Description of the program

This program computes the expected value (mean), variation, and standard deviation of a discrete probability distribution. The user enters the discrete values X and corresponding probabilities P(X). The X values cannot be repeated and the probability values P(X) must be between 0 and 1 inclusive and must add up to 1 within a round-off error of 0.002.

Input:

Number of X values N; N between 1 and 50
Discrete values; X
Corresponding probabilities, P(X); P(X) between 0 and 1
  At the sound of the beep, check your data and record the entry number of any error. For instance, if the error occurred at the data entry 7 out of 10, note the value 7 as the entry number.
Data correction option; If you made an error, use this option with the entry number to correct X or P(X).

II. Sample run

If two fair dice are tossed, the probability for the possible sums shown on the dice are

Sum X	2	3	4	5	6	7	8	9	10	11	12
P(X)	.028	.056	.083	.111	.139	.166	.139	.111	.083	.056	.028

Find the expected sum if you roll two dice. Find the variance and standard deviation of the X distribution

Select the program Expected Value and Standard Deviation of Discrete Probability Distributions.

Next, enter the value 11 for N, since we have 11 different possible sums when we toss two dice. Click on OK. Begin entering your data.

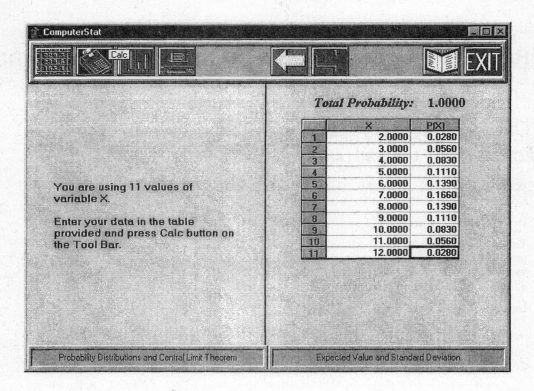

Click on the Calculate button.

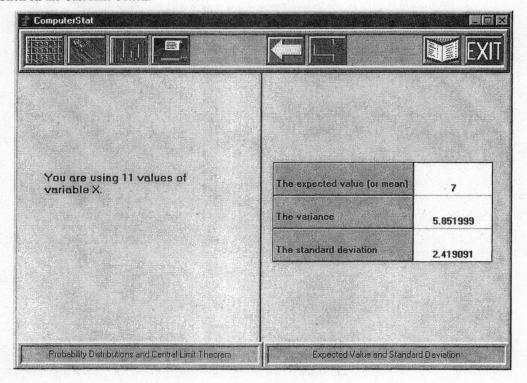

LAB ACTIVITIES FOR DISCRETE PROBABILITY DISTRIBUTIONS

1. Hold Insurance has calculated the following probabilities for claims on a new $15,000 vehicle for one year, if the vehicle is driven by a single male in his twenties.

$ Claim, X	0	1000	5000	10,000	15,000
$P(X)$	0.63	0.24	0.10	0.02	0.01

Find the expected value of a claim in one year. What is the standard deviation of the claim distribution? What should the annual premium be to include $200 in overhead and profit as well as the expected claim?

2. Quick Dental Services claims that the probability distribution for waiting times between the actual appointment time and the time the patient sees the dentists is (in minutes)

Waiting time, X	5	10	15	20	25	30	35	40	45
$P(X)$	0.43	0.28	0.11	0.05	0.04	0.03	0.03	0.02	0.01

Find the expected waiting time and the standard deviation of the waiting time distribution. Based on this information, what would be a reasonable waiting time to advertise?

3. Suppose the possible outcomes of an experiment are labeled 1 through 5, and the reported probabilities of each outcome are as follows.

X	1	2	3	4	5
$P(X)$	0.30	0.25	0.15	0.10	0.15

Enter these values in the program. Why is the probability distribution given above not correct? Do you suppose there is another unreported outcome?

BINOMIAL COEFFICIENTS AND PROBABILITY DISTRIBUTION (SECTIONS 6.2 AND 6.3 OF *UNDERSTANDING BASIC STATISTICS*)

Main menu selection: Probability Distributions and Central Limit Theorem
Sub-menu selection: Binomial Coefficients and Probability Distribution

I. Description of program

Each trial in a binomial experiment has two possible outcomes: success or failure. In this program the user inputs the number N of binomial trials and the probability of success on any one trial. The program displays a table of binomial coefficients C and binomial probabilities P(R), where R is the number of successes in N trials. The cumulative probability $P(0 <= X <= R)$ is also displayed. Note that R can have integer values from 0 to N inclusive. There is an option to graph the distribution.

Input:

The number of trials, N; N between 1 and 40 inclusive

Probability of success on a single trial, P; P between 0 and 1 inclusive

Output: A table of binomial coefficients C and probabilities P(R) and cumulative probabilities
$P(0 <= X <= R)$ for R success out of N trials. Also a graph and values of μ and σ.

II. Sample Run

A surgeon performs a difficult spinal column operation. The probability of success of the operation is
$P = 0.73$. Ten such operations are scheduled. Find the probability of success for 0 through 10 successes out
of these 10 operations.

Select the program Binomial Coefficients and Probability Distribution. The first screen shows the program
title and gives a brief description. Use the number of trials N = 10, and the probability of success P = 0.73.
Click on the Calculate button.

The outcome is the following table.

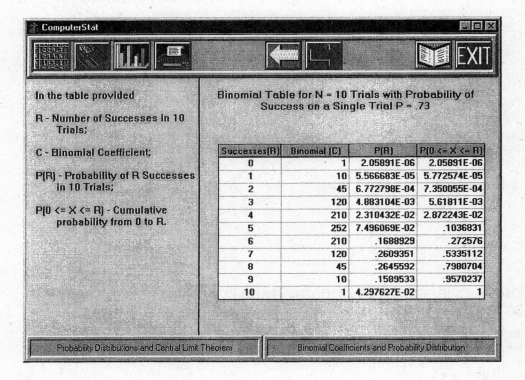

To graph the distribution, click on the Graph button.

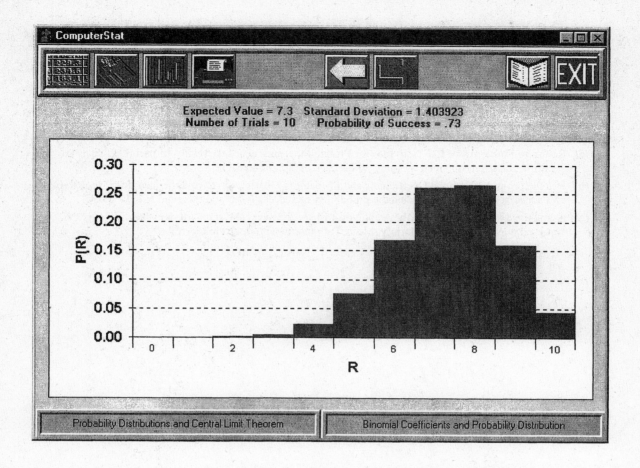

LAB ACTIVITIES USING BINOMIAL COEFFICIENTS AND PROBABILITY DISTRIBUTION

1. You toss a coin 8 times. Call heads success. If the coin is fair, the probability of success P is 0.5. What is the probability of getting exactly 5 heads out of 8 tosses? Of getting exactly 20 heads out of 40 tosses? Of getting 20 heads or fewer out of 40 tosses? Of getting more than 20 heads out of 40? (Hint: Subtract the probability of 20 heads or fewer from 1)?

2. At Pleasant College, the student services office has calculated the probability that a student will get the class schedule he or she wants as 0.83. Five friends register. What is the probability that all 5 of them get their preferred schedule? What is the probability that at least 3 of them get their preferred schedule? Next graph the probability distribution. Is skewed or symmetric? Note the expected value and standard deviation shown above the graph. What is the expected number out of 5 students to get their requested schedule?

3. Some tables for the binomial distribution give values only up to 0.5 for the probability of success. There is a symmetry to the values greater than 0.5 with those values less than 0.5.

   (a) Examine the binomial distribution table in *Understanding Basic Statistics* and see if you can detect the symmetry. Compare the entries for number of successes $n = 3$ and probability of success $p = 0.6$ and $p = 0.4$

   (b) Now use the *ComputerStat* program Binomial Coefficients and Probability Distribution and graph the two distributions with N = 3 and P = 0.6 and P = 0.4. How do the graphs compare? Do the graphs show the same symmetry that the table shows?

(c) Repeat part (b) with N = 10 and $p = 0.5$ and $p = 0.75$.

(d) In general, if P1 = 1 – P2, how will the graphs compare for the same number of trials?

4. In Lab Activities for Probabilities, Problem 4, you experimentally estimated the probability that in a family of three children, all three were girls. We can view this experiment as a binomial experiment. There are two outcomes for each child: girl, which we will call success, and boy, which we will call failure. The probability of success for each child is 0.5. There are three trials, since we are looking at three children.

(a) Return to Problem 4 of Lab Activities for Probabilities and record the results of part (c). This is the experimental result obtained by using 30 families of 3 children each.

(b) Use the program Binomial Coefficients and Probability Distribution to find the theoretical probability of 3 successes out of 3 trials when the probability of success is 0.5 and there are N = 30 trials.

(c) Graph the distribution. Above the graph you will see the mean and standard deviation. What is the expected number of girls in a family selected at random with three children?

CHAPTER 7 NORMAL DISTRIBUTIONS

CONTROL CHARTS
(SECTION 7.1 OF *UNDERSTANDING BASIC STATISTICS*)

Main Menu selection: Probability Distributions and Central Limit Theorem
Sub-menu selection: Control Charts

I. Description of the program

This program asks the user for sequential data (e.g., time series, where data values are separated by equal time units). The sample mean and standard deviation are computed. Then the user is asked for a target mean and target standard deviation to be used in the construction of the control chart. A control chart is printed.

Input:

Option 1: Select a class demonstration.

Use an existing class demonstration data file that is included in the *ComputerStat* data menu. The data values included in the demonstrations are displayed in the Appendix of this guide.

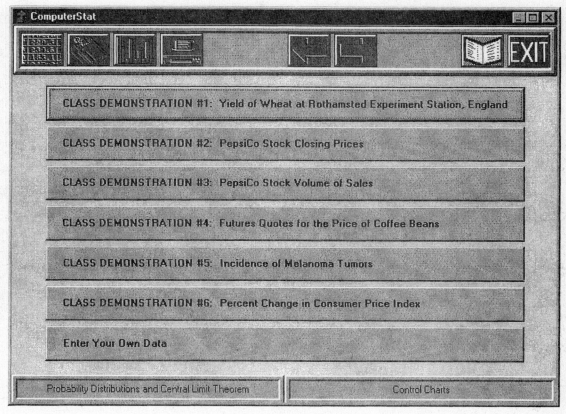

Target values for μ and σ are provided with the data file. However, the user may change these.

Option 2: Enter your own data.

Sample size N; N between 5 and 50 inclusive

Target values for μ and σ

Output:

A display of the data, including target values of μ and σ

II. Sample Run

Let's look at a control chart for the wheat yield in an experimental plot of land at Rothamsted Experiment Station. Select the program CONTROL CHARTS, click on the data button and select Class Demonstration #1. Click on Calculate (the calculator button). Next click on Graph (the button showing a graph). The control chart follows.

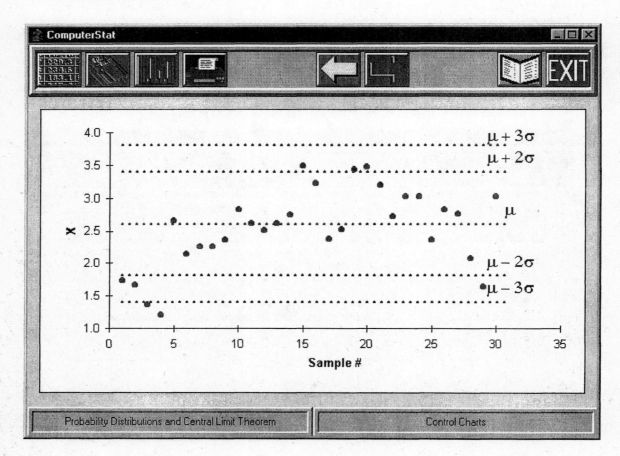

LAB ACTIVITIES FOR CONTROL CHARTS

Use the *ComputerStat* program Control Charts to make control charts for each of the class demonstration options included with the program. Determine if any out-of-control signals are present (see Section 6.1 of *Understanding Basic Statistics*).

(1) Class Demonstration #1: Yield of Wheat at Rothamsted Experiment Station, England

(2) Class Demonstration #2: Pepsico Stock Closing Prices

(3) Class Demonstration #3: Pepsico Stock Volume of Sales

(4) Class Demonstration #4: Future Quotes for the Price of Coffee Beans

(5) Class Demonstration #5: Incidence of Melanoma Tumors

(6) Class Demonstration #6: Percent Change in Consumer Price Index

CHAPTER 8 INTRODUCTION TO SAMPLING DISTRIBUTIONS

CENTRAL LIMIT THEOREM
(SECTION 8.2 OF *UNDERSTANDING BASIC STATISTICS*)

Main Menu selection: Probability Distributions and Central Limit Theorem
Sub-menu selection: Central Limit Theorem Demonstration

I. Description of the program

This program gives a demonstration of the Central Limit Theorem. The Central Limit Theorem says that if x is a random variable with *any* distribution having mean μ and standard deviation σ, then the distribution of means $\bar{x}$ based on random samples of size n is such that for sufficiently large n:

(a) The mean of the $\bar{x}$ distribution is approximately the same as the mean of the x distribution.

(b) The standard deviation of the $\bar{x}$ distribution is approximately $\sigma/\sqrt{n}$.

(c) The $\bar{x}$ distribution is approximately a normal distribution.

Furthermore, as the sample size n becomes larger and larger, the approximations in (a), (b), and (c) become better.

We can use *ComputerStat* to demonstrate the Central Limit Theorem. The computer does not prove the theorem. A proof of the Central Limit Theorem requires advanced mathematics and is beyond the scope of an introductory course. However, we can use the computer to gain a better understanding of the theorem.

To demonstrate the Central Limit Theorem, we need a specific x distribution. One of the simplest is the uniform probability distribution. The normal distribution is the usual bell-shaped curve, but the uniform distribution is the rectangular or box-shaped graph. The two distributions are very different.

The uniform distribution has the property that all subintervals of the same length inside the interval 0 to 9 have the same probability of occurrence no matter where they are located. This means that the uniform distribution on the interval from 0 to 9 could be represented on the computer by selecting random numbers from 0 to 9. Since all numbers from 0 to 9 would be equally likely to be chosen, we say we are dealing with a uniform (equally likely) probability distribution. Note that when we say we are selecting random numbers from 0 to 9, we do not just mean whole numbers or integers; we mean real numbers in decimal form such as 2.413912, and so forth.

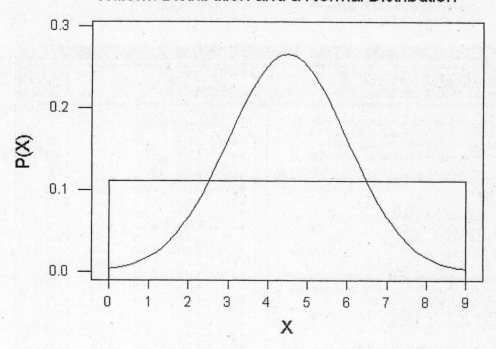

Because the interval from 0 to 9 is 9 units long and because the total area under the probability graph must be 1 (why?), the height of the uniform probability graph must be 1/9. The mean of the uniform distribution on the interval from 0 to 9 is the balance point. Looking at the Figure, it is fairly clear that the mean is 4.5. Using advanced methods of statistics, it can be shown that for the uniform probability distribution x between 0 and 9.

$$\mu = 4.5 \text{ and } \sigma = 3\sqrt{3}/2 \approx 2.598$$

The figure shows us that the uniform x distribution and the normal distribution are quite different. However, using the computer we will construct one hundred sample means $\bar{x}$ from the x distribution using sample sizes of N = 5, 10, 15, 20, and 30. We will see that even though the uniform distribution is very different from the normal distribution, the distribution of sample means $\bar{x}$ comes closer and closer to a bell-shaped normal distribution. This surprising result is predicted by the Central Limit Theorem.

Input:

The means from the 100 samples

The sample mean of the 100 means

The sample standard deviation of the 100 means

A comparison table showing the tally of sample means falling in the specified intervals and the tally predicted by the Central Limit Theorem.

A histogram of the $\bar{x}$ distribution compared to the predicted normal curve.

II. Sample Run

Run the program using sample size N = 40.

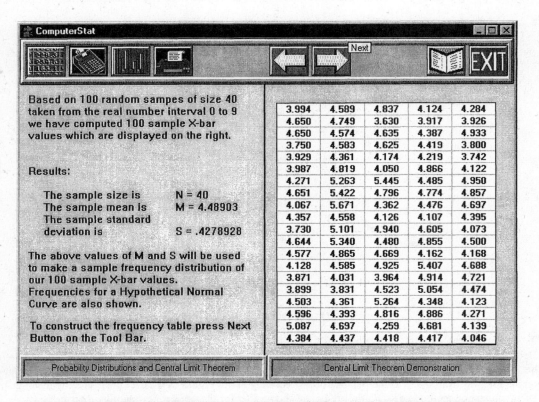

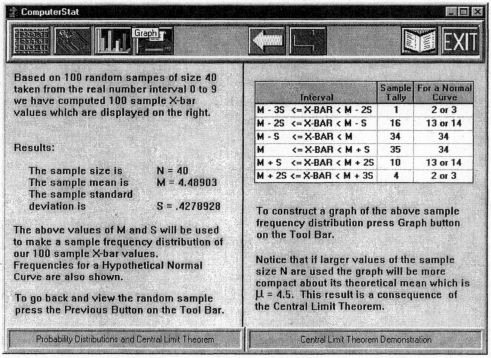

Click on the graph button.

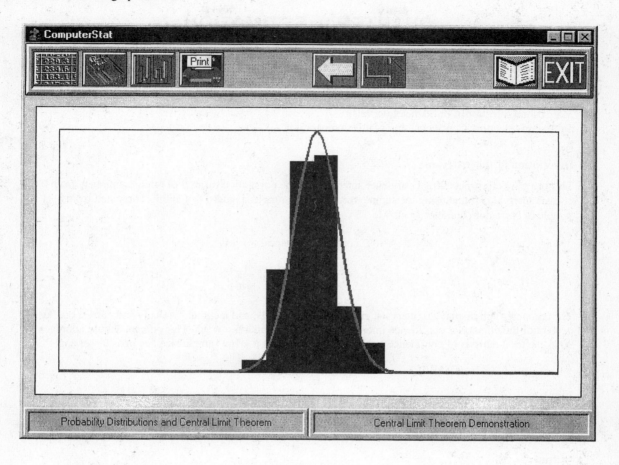

LAB ACTIVITIES FOR CENTRAL LIMIT THEOREM

Using the *ComputerStat* program Central Limit Theorem Demonstration, compare the distribution of sample means taken from 100 samples of size N with the results predicted by the Central Limit Theorem. Look at the results when N is 2, 5, 10, 20, 30, 40, and 50. In each case compare the mean of the $\bar{x}$ distribution with the predicted mean. Compare the tallies of values falling in the specified intervals with those predicted by the Central Limit Theorem. Finally, look at the graphs. Notice what happens as the sample size becomes larger and larger. In general, can we say that the larger the sample size, the narrower the $\bar{x}$ distribution?

CHAPTER 9 ESTIMATION

CONFIDENCE INTERVAL DEMONSTRATION
(SECTION 9.1 OF *UNDERSTANDING BASIC STATISTICS*)

Main Menu selection: Confidence Intervals
Sub-menu selection: Confidence Interval Demonstration

I. Description of the program

This program constructs 90% confidence intervals for the population mean μ of random numbers from 0 to 1. Each interval is found using the sample mean $\overline{x}$ and sample standard deviation s computed from a sample of N random numbers from 0 to 1.

A 90% confidence interval for μ is

$$\overline{x} - 1.645\frac{s}{\sqrt{n}} \text{ to } \overline{x} + 1.645\frac{s}{\sqrt{n}}$$

The computer will display 20 intervals, graphs of the intervals, and a cumulative tally and corresponding percent of the number of confidence intervals that actually contain $\mu = 0.5$. This percent should approach 90 as the total number of confidence intervals we examine of a given sample size becomes larger and larger.

Input:

Sample size N; N between 30 and 500 inclusive

Output:

Table showing sample means, standard deviations and 90% confidence intervals for 20 different random samples. The table also indicates whether or not the confidence interval contains the population mean $\mu = 0.5$.

A graphical display of the 20 confidence intervals described in the table.

Summary indicating the number of confidence intervals of the same sample size, and percent which contain the population mean μ.

II. Sample Run

Generate twenty 90% confidence intervals from samples of size N = 35. Recall that our population consists of the numbers between 0 and 1, so the population mean = 0.5. Enter the value of 35 for N and click on the calculate button.

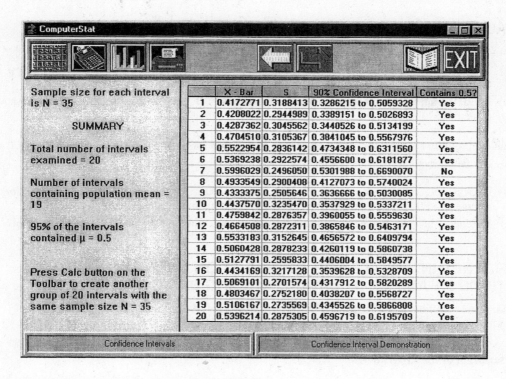

The table shows:

	X - Bar	S	90% Confidence Interval	Contains 0.5?
1	0.4172771	0.3188413	0.3286215 to 0.5059328	Yes
2	0.4208022	0.2944989	0.3389151 to 0.5026893	Yes
3	0.4287362	0.3045562	0.3440526 to 0.5134199	Yes
4	0.4704510	0.3105367	0.3841045 to 0.5567976	Yes
5	0.5522954	0.2836142	0.4734348 to 0.6311560	Yes
6	0.5369238	0.2922574	0.4556600 to 0.6181877	Yes
7	0.5996029	0.2496050	0.5301988 to 0.6690070	No
8	0.4933549	0.2900408	0.4127073 to 0.5740024	Yes
9	0.4333375	0.2505646	0.3636666 to 0.5030085	Yes
10	0.4437570	0.3235470	0.3537929 to 0.5337211	Yes
11	0.4759842	0.2876357	0.3960055 to 0.5559630	Yes
12	0.4664508	0.2872311	0.3865846 to 0.5463171	Yes
13	0.5533183	0.3152645	0.4656572 to 0.6409794	Yes
14	0.5060428	0.2878233	0.4260119 to 0.5860738	Yes
15	0.5127791	0.2595833	0.4406004 to 0.5849577	Yes
16	0.4434169	0.3217128	0.3539628 to 0.5328709	Yes
17	0.5069101	0.2701574	0.4317912 to 0.5820289	Yes
18	0.4803467	0.2752180	0.4038207 to 0.5568727	Yes
19	0.5106167	0.2735569	0.4345526 to 0.5866808	Yes
20	0.5396214	0.2875305	0.4596719 to 0.6195709	Yes

Sample size for each interval is N = 35

SUMMARY

Total number of intervals examined = 20

Number of intervals containing population mean = 19

95% of the intervals contained $\mu = 0.5$

Press Calc button on the Toolbar to create another group of 20 intervals with the same sample size N = 35

Confidence Intervals   Confidence Interval Demonstration

Click on the Graph button to see the intervals displayed.

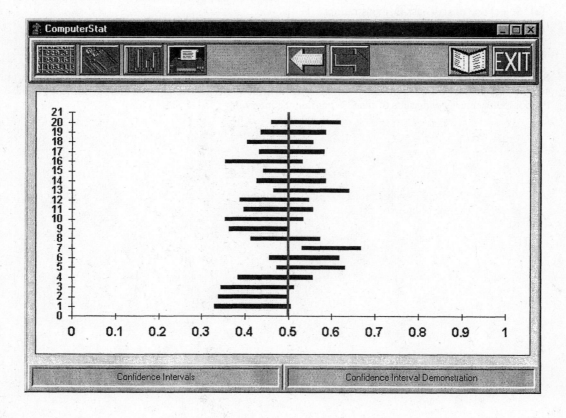

LAB ACTIVITIES FOR CONFIDENCE INTERVAL DEMONSTRATION

1. Run the program with sample of size 35 a total of 5 times to obtain 100 confidence intervals.

    (a) Are each of the intervals the same? Why or why not? Note that the sample means and standard deviations of all samples are slightly different.

    (b) Do all the intervals contain the population mean of 0.5? In the long run, about what percent should contain 0.5?

2. Run the program for intervals from sample size 30 and then for intervals with sample size 50.

    (a) Do the intervals seem to be shorter when we use a sample of size 50?

    (b) Do you predict that the intervals for samples of size 100 will be shorter or longer than those for sample size 50? Run the program with N = 100 to verify your answer.

    (c) In general, how can you narrow the confidence interval without changing the confidence level?

CONFIDENCE INTERVALS FOR A POPULATION MEAN
(SECTIONS 9.1 AND 9.2 OF *UNDERSTANDING BASIC STATISTICS*)

Main menu selection: Confidence Intervals
Sub-menu selection: Confidence Intervals for a Population Mean μ

I.   Description of the Program

This program finds confidence intervals for a population mean μ. The user may

(a) Select one of six Class Demonstrations

(b) Enter sample raw data

(c) Enter processed data; the sample mean $\overline{x}$ and sample standard deviation s

Small samples (sample size less than 30) use critical values from the inverse Student's t distribution. Large samples use critical values from the inverse normal distribution. The user inputs a confidence level C and the computer returns endpoints of the C% confidence interval.

Input:

Option 1: Select a classroom demonstration:

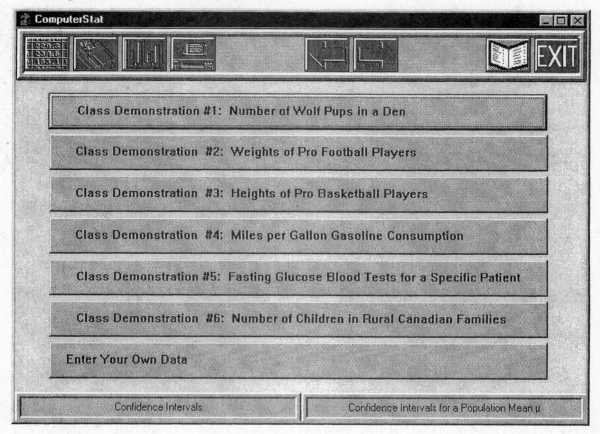

Option 2:

    (a) Enter raw data.

    (b) Enter processed data: Specify sample size N, mean of the data x and sample standard deviation s

Sample size, N; for raw data N must be between 2 and 100 inclusive

Confidence level, C: C as a percent greater than 1 and less than 100

Output:

Computer tells you if you have a large or small sample

Information summary: confidence level, sample size, sample mean, sample standard deviation

Critical value of Z_0 for large samples

Critical value of t_0 for small samples

The C% confidence interval for the population mean μ

II. Sample Run

Start the program and select Classroom Demonstration #1: Number Of Wolf Pups In A Den. Use a 90% confidence level.

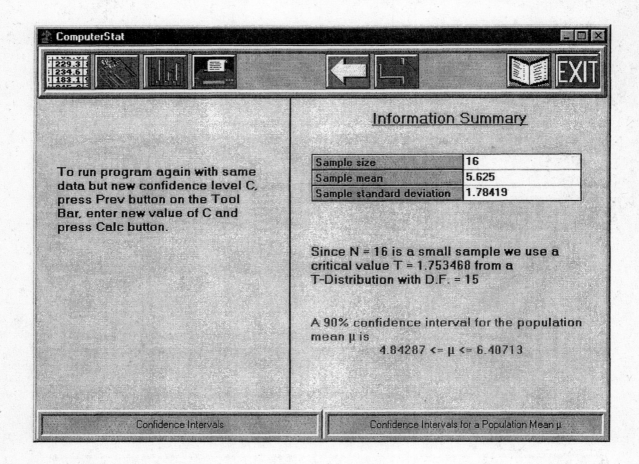

LAB ACTIVITIES FOR CONFIDENCE INTERVALS FOR A POPULATION MEAN

1. Market Survey was hired to do a study for a new soft drink, Refresh. A random sample of 20 people were given a can of Refresh and asked to rate it for taste on a scale of 1 to 10 (with 10 being the highest rating). The ratings were

5	8	3	7	5	9	10	6	6	2
9	2	1	8	10	2	5	1	4	7

   Find an 85% confidence interval for the population mean rating of Refresh.

2. Select Class Demonstration #3: Heights of Pro Basketball Players.
   (a) Find a 99% confidence interval for the population mean height.
   (b) Find a 95% confidence interval for the population mean height.
   (c) Find a 90% confidence interval for the population mean height.
   (d) Find an 85% confidence interval for the population mean height.
   (e) What do you notice about the length of the confidence interval as the confidence level goes down? If you used a confidence level of 80%, would you expect the confidence interval to be longer or shorter than that of 85%? Run the program again to verify your answer.

3. Under the option to enter your own data, you can choose to enter only the sample size N, the sample mean x and the sample standard deviation s for a data set. Select this option. Suppose you read a newspaper article in which a random sample of U.S. income tax returns was reported to have an average time for a refund to be issued of 45 days, with sample standard deviation of 8 days.

   (a) If the sample size N = 20, find a 95% confidence interval for the population mean number of days required to issue a refund.

   (b) If the sample size N = 40, find a 95% confidence interval for the population mean number of days required to issue a refund.

   (c) If the sample size N = 60, find a 95% confidence interval for the population mean number of days required to issue a refund.

   (d) If the sample size N =100, find a 95% confidence interval for the population mean number of days required to issue a refund.

   (e) What do you notice about the length of the confidence interval as the sample size increases? If you used a sample size of N = 500, would you expect the confidence interval to be longer or shorter than when you use a sample size of 100? Run the program again with N = 500 to verify your answer.

CONFIDENCE INTERVALS FOR THE PROBABILITY OF SUCCESS p IN A BINOMIAL DISTRIBUTION (SECTION 9.3 OF *UNDERSTANDING BASIC STATISTICS*)

Main menu selection: Confidence Intervals
Sub-menu selection: Confidence Intervals for P the Probability of Success in a Binomial Distribution

l. Description of the program

This program finds confidence intervals for P, the probability of success on a single trial of a binomial distribution. We assume that the number of trials N is large enough to permit the use of a normal approximation to the binomial distribution. Empirical studies show that the normal approximation is reasonable if both NP and N(1 – P) are both greater than 5. The computer checks that these conditions are met.

Input:

The percent confidence level, C; C as a percent between 1 and 100

Number of trials, N; N greater than or equal 12

Number of successes out of N trials, R; R between 0 and N inclusive

Output:

Information summary listing percent confidence level C, number of trials, sample approximation for P, comment indicating if a normal approximation to the binomial is appropriate

The confidence interval for P

II. Sample Run

The public television station BPBS wants to find the percent of its viewing population who give donations to the station. A random sample of 300 viewers were surveyed, and it was found that 123 made contributions to the station. Find a 95% confidence interval for the probability that a viewer of BPBS selected at random contributes to the station.

Select the program. The program begins with its title and brief description. Since we have binomial trials, this program is appropriate.

Enter the value 95 for the percent confidence level. Use the value N = 300, since there are 300 viewers in the sample. The number of successes is R = 123. The output follows.

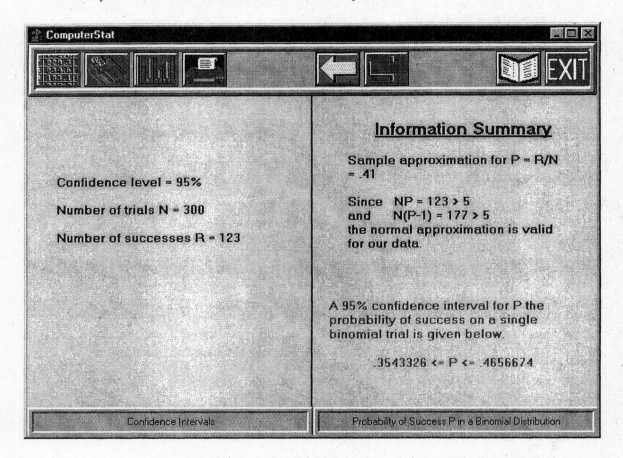

LAB ACTIVITIES FOR CONFIDENCE INTERVALS FOR THE PROBABILITY OF SUCCESS *p* IN A BINOMIAL DISTRIBUTION

1. Many types of error will cause a computer program to terminate or give incorrect results. One type of error is punctuation. For instance, if a comma is inserted in the wrong place, the program might not run. A study of programs written by students in a beginning programming course showed that 75 out of 300 errors selected at random were punctuation errors. Find a 99% confidence interval for the proportion of errors made by beginning programming students that are punctuation errors. Next, find a 90% confidence interval. Is this interval longer or shorter?

2. Sam decided to do a statistics project to determine a 90% confidence interval for the probability that a student at West Plains College eats lunch in the school cafeteria. He surveyed a random sample of 12 students and found that 9 ate lunch in the cafeteria. Can Sam use the program to find a confidence interval for the population proportion of students eating in the cafeteria? Why or why not? Try the program with N = 12 and R = 9. What happens? What should Sam do to complete his project?

CHAPTER 10 HYPOTHESIS TESTING

TESTING A SINGLE POPULATION MEAN
(SECTIONS 10.1–10.4 OF *UNDERSTANDING BASIC STATISTICS*)

Main Menu selection: Hypothesis Testing
Sub-menu selection: Testing a Single Population Mean μ

I.   Description of the program

Let μ be the population mean of a distribution of x values. We will assume that the *x* distribution is one for which both the population mean and standard deviation are unknown. In the case where we have only a small sample, we also must assume that the x distribution is approximately a normal distribution. This program tests the null hypothesis

$$H0: \mu = K$$

against the alternate hypothesis, which may be

$$H1: \mu < K \text{ (left-tailed test)}$$
$$H1: \mu > K \text{ (right-tailed test)}$$
$$H1: \mu <> K \text{ (two-tailed test)}$$

The program performs the hypothesis test using P as well as the critical region and sample test statistic.

Input:

Option 1: Select a class demonstration:

(1) Class Demonstration #1: Number of Wolf Pups in a Den

(2) Class Demonstration #2: Weights of Pro Football Players

(3) Class Demonstration #3: Heights of Pro Basketball Players

(4) Class Demonstration #4: Miles per Gallon Gasoline Consumption

(5) Class Demonstration #5: Fasting Glucose Blood Tests for a Specific Patient

(6) Class Demonstration #6: Number of Children in Rural Canadian Families

Option 2: Enter your own data.

(a) Enter raw data

(b) Enter processed data, that is: sample size, sample mean, sample standard deviation

The constant K in the null hypothesis H0: $\mu = K$

Type of test:

$$H1: \mu < K \text{ (left-tail)}$$

$$H1: \mu > K \text{ (right-tail)}$$

$$H1: \mu <> K \text{ (two-tail)}$$

Level of significance, alpha; alpha between 0 and 0.5.

Output:

Information summary table listing sample size, sample mean, sample standard deviation, level of significance, null hypothesis, alternate hypothesis, P value test conclusion with mention of distribution used, T or Z value corresponding to the sample test statistic $\overline{x}$, location of sample test statistic inside or outside the critical region, and P value compared to the level of significance alpha.

II. Sample Run

Start the program and select Class Demonstration #3 HEIGHTS OF PRO BASKETBALL PLAYERS. Notice that the level of significance and hypothesis are already given. These may be changed later. At the alpha = 0.05 level of significance, do the data support the conclusion that the average height of the players is different from 6.5 feet?

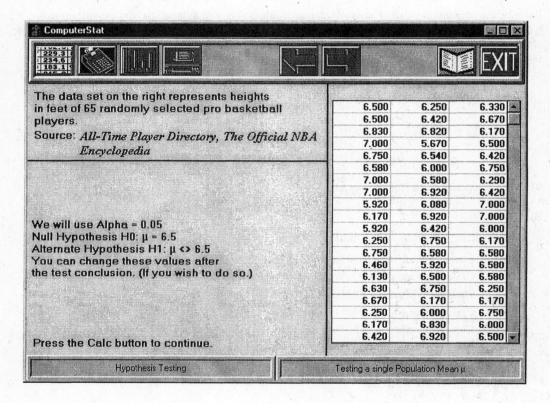

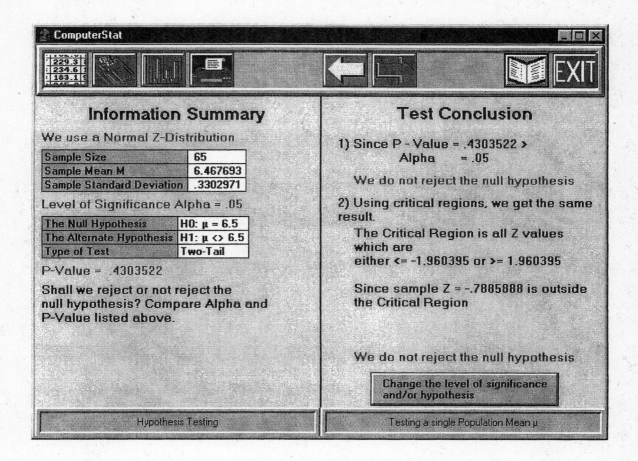

LAB ACTIVITIES FOR TESTING A SINGLE POPULATION MEAN

1. Select Class Demonstration #3 regarding the heights of pro basketball players. In the conclusion window, select the option to change the hypothesis and/or level of significance.

   This time test the hypothesis that the average height of the players is greater than 6.2 feet at the 1% level of significance. To do this, change the value of K to 6.2 and the alternate hypothesis to a right-tailed test. Change the level of significance to 0.01.

2. In this problem we will see how the test conclusion is possibly affected by a change in the level of significance.

   Teachers for Excellence is interested in the attention span of students in grades 1 and 2, now as compared to 20 years ago. They believe it has decreased. Studies done 20 years ago indicate that the attention span of children in grades 1 and 2 was 15 minutes. A study sponsored by Teachers for Excellence involved a random sample of 20 students in grades 1 and 2. The average attention span of these students was (in minutes) $\bar{x} = 14.2$ with standard deviation $s = 1.5$.

   (a) Use the program to conduct the hypothesis test using $\alpha = 0.05$ and a left-tailed test. What is the test conclusion? What is the P value?

   (b) Use the program to conduct the hypothesis test using $\alpha = 0.01$ and a left-tailed test. What is the test conclusion? How could you have predicted this result by looking at the P value from part (a)? Is the P value for this part the same as it was for part (a)?

3. In this problem, let's explore the effect that sample size has on the process of testing a mean. Run the program with the hypotheses $H_0: M = 200$, $H_1: \mu > 200$, $\alpha = 0.05$, $\bar{x} = 210$ and $s = 40$.

   (a) Use the sample size N = 30. Note the *P* value and Z score of the sample test statistic and test conclusion.

   (b) Use the sample size N = 50. Note the *P* value and Z score of the sample test statistic and test conclusion.

   (c) Use the sample size N = 100. Note the *P* value and Z score of the sample test statistic and test conclusion.

   (d) In general, if your sample statistic is close to the proposed population mean specified in H_0, and you want to reject H_0, would you use a smaller or a larger sample size?

TESTING A SINGLE PROPORTION
(SECTIONS 10.5 OF *UNDERSTANDING BASIC STATISTICS*)

   Main menu selection: Hypothesis Testing
   Sub-menu selections: Testing a Single Proportion P

I.   Description of the Programs

   This program is similar to the program Testing a Single Mean μ in the sense that you enter the target value K for the null hypothesis, select the alternate hypothesis, and specify the level of significance alpha. You also enter the number of trials and the number of successes. The program displays an information summary, the Z or T value corresponding to the sample test statistic, the P value corresponding to the sample test statistic, and the test conclusion.

   Input: The number of trials
   The number of successes
   The constant K in the null hypothesis H0: P = K
   Type of test: H1:P < K (left tail)

   H1:P > K (right tail)

   H1:P <> K (two tail)

   Level of significance, alpha: alpha between 0. and 0.5

   Output:

   Information summary

   Z or T value of the sample test statistic

   P value of the sample test statistic

   Test conclusion

LAB ACTIVITIES FOR TESTS OF A SINGLE PROPORTION

1. Jones Computer Security is testing a new security device which is believed to decrease the incidence of computer "break-ins." Without this device, the computer security test team can break security 47% of the time. With the device in place, the test team made 400 attempts and were successful 82 times. Select an appropriate program from the HYPOTHESIS TESTING menu and test the claim that the device reduces the proportion of successful break-ins. Use alpha = 0.05 and note the P value. Does the test conclusion change for alpha = 0.01?

2. The programs of *ComputerStat* offer a tremendous aid in performing the computations required for hypothesis testing. However, the user must identify which program to use, select the hypotheses, and interpret the test conclusions in terms of the original statement of a problem. Select appropriate programs from *ComputerStat* to do the even-numbered problems of Chapter 10 Review Problems in *Understanding Basic Statistics*.

CHAPTER 11 INFERENCES ABOUT DIFFERENCES

OTHER TESTS OF HYPOTHESIS
(SECTIONS 11.1–11.4 OF *UNDERSTANDING BASIC STATISTICS*)

Main menu selection: Hypothesis Testing
Sub-menu selections: Testing $\mu 1 - \mu 2$ (Dependent Samples)
        (Section 11.1 of *Understanding Basic Statistics*)
      Testing $\mu 1 - \mu 2$ (Independent Samples)
        (Sections 11.2 and 11.3 of *Understanding Basic Statistics*)
      Testing P1 – P2 (Large Samples)
        (Section 11.4 of *Understanding Basic Statistics*)

I. Description of the Programs

All of these programs are similar to the program Testing a Single Mean μ in the sense that you enter the target value K for the null hypothesis, select the alternate hypothesis, and specify the level of significance alpha. Each program displays an information summary, the Z or T value corresponding to the sample test statistic, the P value corresponding to the sample test statistic, and the test conclusion.

Input: All the programs have options to enter your own data.

Testing $\mu 1 - \mu 2$ (Dependent Samples) has these class demonstrations:

  (1) Class Demonstration #1: Annual Salaries for Male Versus Female Assistant Professors
  (2) Class Demonstration #2: Unemployment Percentage for High School only Versus College
                       Graduates
  (3) Class Demonstration #3: Number of Traditional Navajo Hogans Versus Number Houses on the
                       Navajo Indian Reservation
  (4) Class Demonstration #4: Average Monthly Temperature in Miami Versus Honolulu

Testing MU1-MU2 (Independent Samples) has these Class Demonstrations:

  (1) Class Demonstration #1: Heights of Pro Football Players Versus Heighs of Pro Basketball Players
  (2) Class Demonstration #2: Weights of Pro Football Players Versus Weights of Pro Basketball
                       Players
  (3) Class Demonstration #3: Sepal Width of Iris Versicolor Versus Iris Virginica
  (4) Class Demonstration #4: Number of Cases of Red Fox Rabies in Two Regions of Germany

Output:

Information summary

Z or T value of the sample test statistic

P value of the sample test statistic

Test conclusion

LAB ACTIVITIES USING TESTS ABOUT DIFFERENCES

1. Select TESTING $\mu 1 - \mu 2$ (DEPENDENT SAMPLES) and use one of the class demonstrations.
  (a) Use alpha = 0.01 with a two-tailed test and note the P value and test conclusion.
  (b) Use alpha = 0.05 with a two-tailed test and note the P value and test conclusion.

(c) Does the P value change? Does the sample test statistic change? Does the critical region change? Does the test conclusion change?

2. Select TESTING μ1 – μ2 (INDEPENDENT SAMPLES) and use one of the class demonstrations.

(a) Use alpha = 0.01 with a two-tailed test and note the value of the sample test statistic $\overline{x}_1 - \overline{x}_2$ and test conclusion.

(b) Select the same class demonstration from CONFIDENCE INTERVALS FOR μ1 – μ2 (INDEPENDENT SAMPLES). Find a 99% confidence interval and record the endpoints of the interval.

(c) Is the value of the sample test statistic $\overline{x}_1 - \overline{x}_2$ found in part (a) inside or outside the confidence interval? If it is inside, was the test conclusion of part (a) "Do not Reject H_0"? If it is outside, was the test conclusion "Reject H_0"?

3. Publisher's Survey did a study to see if the proportion of men who read mysteries is different from the proportion of women who read them. A random ample of 402 women showed that 112 read mysteries regularly (at least 6 books per year). A random sample of 365 men showed that 92 read mysteries regularly. Is the proportion of mystery readers different between men and women? Use a 1% level of significance.

(a) Look at the P value of the test conclusion. Jot it down.

(b) Test the hypothesis is that the proportion of women who read mysteries is *greater* than the proportion of men. Use a 1% level significance. Is the P value for a right-tailed test half that of a two-tailed test? If you know the P value for a two-tailed test, can you draw conclusions for a one-tailed test?

4. The programs of *ComputerStat* offer a tremendous aid in performing the computations required for hypothesis testing. However, the user must identify which program to use, select the hypotheses, and interpret the test conclusions in terms of the original statement of a problem. Select appropriate programs from *ComputerStat* to do the even-numbered problems of Chapter 11 Review Problems in *Understanding Basic Statistics*

CONFIDENCE INTERVALS FOR $\mu_1 - \mu_2$ (INDEPENDENT SAMPLES)
CONFIDENCE INTERVALS FOR $p_1 - p_2$ (LARGE SAMPLES)
(SECTIONS 11.2–11.4 OF *UNDERSTANDING BASIC STATISTICS*)

Main menu selection: Confidence Intervals
Sub-menu selection: Confidence Intervals for $\mu1$-$\mu2$ (Independent Samples),
                    Confidence Intervals for P1-P2 (Large Samples)

I. **Description of the Programs**

These programs find C% confidence intervals for the difference of means (independent samples) and difference of proportions respectively. Confidence intervals for difference of means are found for both large and small independent samples. Confidence intervals for difference of proportions assume that the number of trials for each proportion are large enough to justify the use of the normal approximation to the binomial distribution.

Input:

Difference of means

Option 1: Select a class demonstration

Option 2: Enter your own data

(a) Enter the raw data for each distribution

   (b) Enter processed data: Sample size, sample mean, and sample standard deviation for each
       distribution.

   Percent confidence level C; C greater than 1 and less than 100

Difference of proportions

   Number of trials for first distribution, N1; N1 at least 12

   Number of successes for first distribution, R1

   Number of trials for second distribution, N2; N2 at least 12

   Percent confidence level C; C greater than 1 and less than 100

Output:

   Information summary

   The confidence interval for the difference

II. Sample Run (Difference of Means)

Select Class Demonstration #1 to find a 95% confidence interval for the difference of means in heights of
pro football players and pro basketball players. Does the confidence interval indicate that the mean height
of basketball players is different from the mean height of football players?.

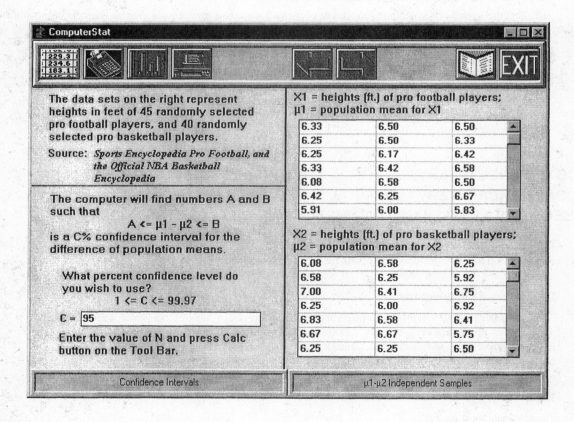

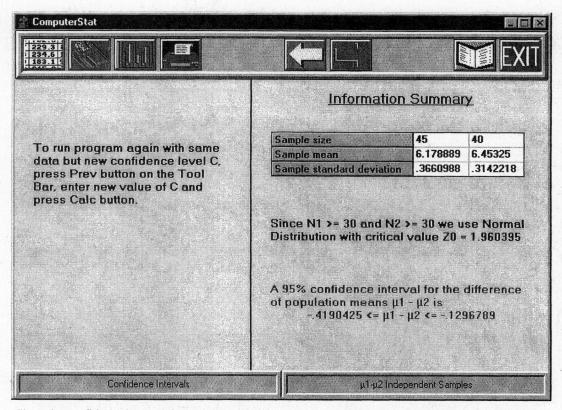

Since the confidence interval does not contain the number zero, it indicates that the mean height of the basketball players is different from the mean height of football players μ_1.

LAB ACTIVITIES FOR CONFIDENCE INTERVALS FOR $\mu_1 - \mu_2$ OR FOR $p_1 - p_2$

1. Run Confidence Intervals for $\mu_1-\mu_2$. Select Class Demonstration #4 – Number of Cases of Red Fox Rabies in Two Regions of Germany.

   (a) Use a confidence level C% = 90. Does the interval indicate that the population mean number of rabies in region 1 is greater than the population mean number of rabies in region 2 at the 90% level? Why or why not?

   (b) Use a confidence level C% = 95. Does the interval indicate that the population mean number of rabies in region 1 is greater than the population mean number of rabies in region 2 at the 95% level? Why or why not?

   (c) Predict whether or not the population mean number of rabies in region 1 is greater than the population mean number of rabies in region 2 at the 99% level. Explain your answer. Verify your answer by running the program with a 99% confidence level.

2. A random sample of 30 police officers working the night shift showed that 23 used at least 5 sick leave days per year. Another random sample of 45 police officers working the day shift showed that 26 used at least 5 sick leave days per year. Use the program Confidence Intervals for P1-P2 to find the 90% confidence interval for the difference of population proportions of police officers working the two shifts using at least 5 sick leave days per year. At the 90% level, does it seem that the proportions are different? Does there seem to be a difference in proportions at the 99% level? Why or why not

CHAPTER 12 ADDITIONAL TOPICS USING INFERENCES

CHI-SQUARE TEST OF INDEPENDENCE
(SECTION 12.1 OF *UNDERSTANDING BASIC STATISTICS*)

Main Menu selection: Hypothesis Testing

Sub-menu selection: χ^2 Test for Independence

I. Description of the program

This program uses the chi-square distribution to test for statistical independence of variables. The hypotheses are

H_0: The variables are independent.

H_0: The variables are not independent.

The computer asks the user to input the number of rows (R) and the number of columns (C) in the contingency table, the observed frequencies for each cell, and the level of significance, alpha, for the tests. The type of test will always be a right-tailed test on a chi-square distribution with degrees of freedom d.f. = (R − 1)(C − 1).

The expected frequency for each cell in the contingency table is calculated. If the expected frequency in any cell is less than five, a message stating that the sample size is too small will be given and new data will be requested. The output includes a completed contingency table with the observed and expected frequencies. The sample chi-square statistic is given, as well as the corresponding P value. The test is concluded by comparing the P value to the level of significance.

Input:

Option 1: Select a class demonstration:

(1) Class Demonstration #1: Salary vs. Job Satisfaction

(2) Class Demonstration #2: Anxiety Level vs. Need To Succeed

(3) Class Demonstration #3: Age vs. Movie Preference

(4) Class Demonstration #4: Number Of Library Contributors vs. Ethnic Group

Option 2: Enter your own data with data correction option.

Number of rows in the contingency table, R; R between 2 and 10 inclusive

Number of columns in the contingency table, C; C between 2 and 10 inclusive

Level of significance, alpha

Output:

Completed contingency table with the row and column number of each cell, the observed frequency in each cell, and the expected frequency of each cell

Information summary showing the null hypothesis, alternate hypothesis, total sample size, number of rows in the contingency table, number of columns in the contingency table, level of significance, sample chi-square value, P value of the sample statistic

Test conclusion based on comparison of the P value to α

II. Sample Run

A computer programming aptitude test has been developed for high school seniors. The test designers claim that scores on the test are independent of the type of school the student attends: rural, suburban, urban. A study involving a random sample of students from each of these types of institutions yielded the following information, where aptitude scores range from 200 to 500, with 500 indicating the greatest aptitude and 200 the least. The entry in each cell is the observed number of students achieving the indicated score on the test.

School Type

Score	Rural	Suburban	Urban
200–299	33	65	82
300–399	45	79	95
400–500	21	47	63

Using the program χ^2 Test of Independence, test the claim that the aptitude test scores are independent of the type of school attended, at the 0.05 level of significance.

We have R = 3 rows and C = 3 columns. Enter these numbers when requested. The level of significance is alpha = 0.05. Enter the data of the contingency table of rows. The results are

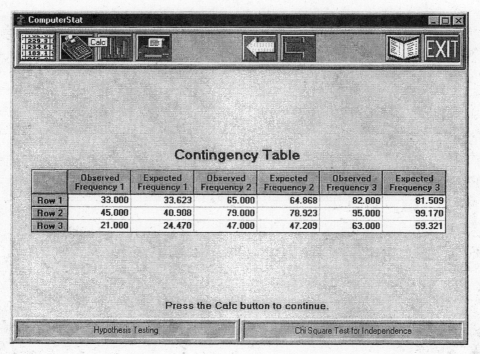

Contingency Table

	Observed Frequency 1	Expected Frequency 1	Observed Frequency 2	Expected Frequency 2	Observed Frequency 3	Expected Frequency 3
Row 1	33.000	33.623	65.000	64.868	82.000	81.509
Row 2	45.000	40.908	79.000	78.923	95.000	99.170
Row 3	21.000	24.470	47.000	47.209	63.000	59.321

Press the Calc button to continue.

Hypothesis Testing	Chi Square Test for Independence

Click on the calculator button to conclude the test.

LAB ACTIVITIES FOR CHI-SQUARE TEST OF INDEPENDENCE

1. We Care Auto Insurance had its staff of actuaries conduct a study to see if vehicle type and loss claim are independent. A random sample of auto claims over six months gives the information in the contingency table.

Total Loss Claims per Year per Vehicle

Type of vehicle	$0–999	$1000–2999	$3000–5999	$6000+
Sports car	20	10	16	8
Truck	16	25	33	9
Family	40	68	17	7
Compact	52	73	48	12

Test the claim that car type and loss claim are independent. Use $\alpha = 0.05$.

2. An education specialist is interested in comparing three methods of instruction:

      SL – standard lecture with discussion

      TV – video taped lectures with no discussion

      IM – individualized method with reading assignments and tutoring, but no lectures.

The specialist conducted a study of these three methods to see if they are independent. A course was taught using each of the three methods and a standard final exam was given at the end. Students were put into the different method sections at random. The course type and test results are shown in the next contingency table.

Final Exam Score

Course Type	< 60	60–69	70–79	80–89	90–100
SL	10	4	70	31	25
TV	8	3	62	27	23
IM	7	2	58	25	22

Test the claim that the instruction method and final exam test scores are independent, using $\alpha = 0.01$.

3. Select one of the class demonstrations and use *ComputerStat* to do the corresponding Section 12.1 problem of *Understanding Basic Statistics*.

CHI-SQUARE TEST FOR GOODNESS OF FIT
(SECTION 12.2 OF *UNDERSTANDING BASIC STATISTICS*)

Main menu selection: Hypothesis Testing

Sub-menu selection: χ^2 Test for Goodness of Fit

I. Description of the Program

This program uses the chi-square distribution to test for goodness of fit. The hypotheses are

H_0: The population from which sample measurements are taken fits a given distribution.

H_1: The population from which sample measurements are taken does not fit the given distribution.

The user inputs the number of items (that is, categories) in the sample distribution, the total sample size, the observed sample frequency for each item, and the expected percent of total sample size for each item. The level of significance is also requested.

The type of test will always be a right-tailed test on a chi-square distribution with degrees of freedom d.f. = (Number of items – 1). The computer will calculate the expected frequency for each item. If the expected frequency for any item is less than five, a message stating that the sample size is to small will be given.

The test conclusion is based on the P value of the sample chi-square statistic.

Input:

Option 1: Select a class demonstration:

   (1) Class Demonstration #1: Library Circulation Distribution
   (2) Class Demonstration #2: Electrical Power Distribution
   (3) Class Demonstration #3: Hospital Patient Distribution
   (4) Class Demonstration #4: Ethnic Distibution

Option 2: Enter your own data.

Number of items in distribution; value between 2 and 50 inclusive

Observed frequency for each item, O(I)

Expected percent of total sample size per item, E1(I)

Output:

Expected frequency per item, E(I)

Information summary

Sample chi-square value

P value

II. Sample Run

Wenland College just changed the length of its summer session from 10 weeks to 8 weeks. The director of the self-paced individualized instruction center wants to see if the grade distribution for courses offered in the center has changed with the shortened summer session. Past records give the grade distribution for courses conducted in a 10-week session. From these figures, the expected percent of the sample size can be computed for each item. A random sample of 216 students showed the grade distribution for the 8-week summer session.

Item Grade	Observed Frequency From sample (8 weeks)	Expected Frequency % (10 week session)
A	28	15
B	73	40
C	40	15
D	8	3
Incomplete	49	18
Withdrawal	18	9

Use χ^2 Goodness of Fit to see if the grade distribution for the 8-week summer session is different from that of the 10-week session at the 0.05 level of significance.

The sample size is n = 216. The number of items in the sample distribution is 6. Enter that value when requested. The level of significance is 0.05. Next enter the observed frequencies and expected percent of total sample size. The computer immediately computes the expected frequency for that item. A data summary is provided, followed by the information summary.

After entering the data, click on the Calculator key.

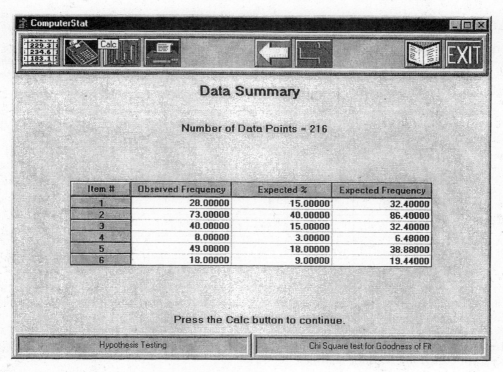

Data Summary

Number of Data Points = 216

Item #	Observed Frequency	Expected %	Expected Frequency
1	28.00000	15.00000	32.40000
2	73.00000	40.00000	86.40000
3	40.00000	15.00000	32.40000
4	8.00000	3.00000	6.48000
5	49.00000	18.00000	38.88000
6	18.00000	9.00000	19.44000

Press the Calc button to continue.

Hypothesis Testing	Chi Square test for Goodness of Fit

Click on the calculator key again to conclude the test.

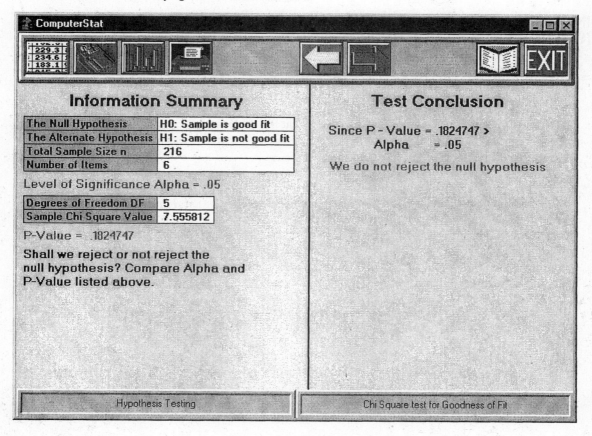

Information Summary

The Null Hypothesis	H0: Sample is good fit
The Alternate Hypothesis	H1: Sample is not good fit
Total Sample Size n	216
Number of Items	6

Level of Significance Alpha = .05

Degrees of Freedom DF	5
Sample Chi Square Value	7.555812

P-Value = .1824747

Shall we reject or not reject the null hypothesis? Compare Alpha and P-Value listed above.

Test Conclusion

Since P - Value = .1824747 >
Alpha   = .05

We do not reject the null hypothesis

Hypothesis Testing	Chi Square test for Goodness of Fit

LAB ACTIVITIES FOR CHI-SQUARE TEST FOR GOODNESS OF FIT

1. Market Survey is doing a study of moderately priced restaurants (dinner for less than $15) according to type. Five years ago they had done a similar study in the Midwest. The new study is being conducted in the same area of the country to determine if the distribution of restaurants has changed. A random sample of 129 restaurants in the Midwest was used.

Item	Observed Frequency	Expected Frequency %
American	51	47
Chinese food	14	6
Mexican food	15	10
Italian food	27	22
Other ethnic food	14	10
Salad bar	8	5

At the 5% level of significance, test the claim that there has been a shift in the distribution of restaurants according to food type.

2. Select one of the class demonstrations and use *Computer Stat* to do the corresponding Section 12.2 problem of *Understanding Basic Statistics*.

INFERENCES RELATING TO LINEAR REGRESSION
(SECTIONS 12.4–12.5 OF *UNDERSTANDING BASIC STATISTICS*)

Main Menu selection: Linear Regression and Correlation
Sub-menu selection: Linear Regression and Testing the Correlation Coefficient Rho
Main Menu selection: Hypothesis Testing
Sub-menu selection: Linear Regression and Testing the Correlation Coefficient Rho
The same program that we used in Chapter 4 for Linear Regression and Correlation include the value of the standard error of estimate (S_e) in the information summary, the option to construct a confidence interval, and the option to test the population correlation coefficient ρ.

I. **Description of the program**

The program Linear Regression and Correlation performs a number of function related to linear regression and correlation. The user inputs a random sample of ordered pairs of X and Y values. Then the computer returns the sample means, sample standard deviations of X and Y values; largest and smallest X value; standard error of estimate; equation of least-squares line; the correlation coefficient and the coefficient of determination.

Next the user has several options.

1. Predict Y values from X values, and when the X value is in the proper range, compute C% confidence intervals for Y values.

2. Test the correlation coefficient. If we let RHO represent the population correlation coefficient, then the null hypothesis is H0: RHO = 0. The alternate hypothesis may be one of the following.

$$H1: RHO < 0 \text{ (left-tailed test)}$$
$$H1: RHO > 0 \text{ (right-tailed test)}$$
$$H1: RHO <> 0 \text{ (two tailed-test)}$$

3. Graph data points (X, Y) to create a scatter diagram. Then the user can see five points on the least-squares line.

Input:

Option 1: Select a class demonstration:

(1) Class Demonstration #1: List Price vs. Best Price For A New GMC Pickup Truck

(2) Class Demonstration #2: Cricket Chirps vs. Temperature

(3) Class Demonstration #3: Diameter of Sand Granules vs. Slope on a Naturally Occurring Ocean Beach

(4) Class Demonstration #4: National Unemployment Rate Male vs. Female

Option 2:

Value of X for which you want to predict Y

Confidence level C for confidence interval of Y values

Alternate hypothesis for testing RHO

Level of significance for testing RHO

Output:

Information summary table including the least squares line, S_e, R, R^2

For a given X value, the predicted Y value and a C% confidence interval

Test summary for testing RHO

Scatter diagram and least-squares line

II. Sample Run

Begin the program Linear Regression and Testing the Correlation Coefficient Rho. Click on the Data button. Select Class Demonstration #2) Cricket Chirps vs. Temperature. Use the program to explore the relationship between cricket chirp frequency and temperature.

After you select the class demonstration, the data is shown followed by the information summary.

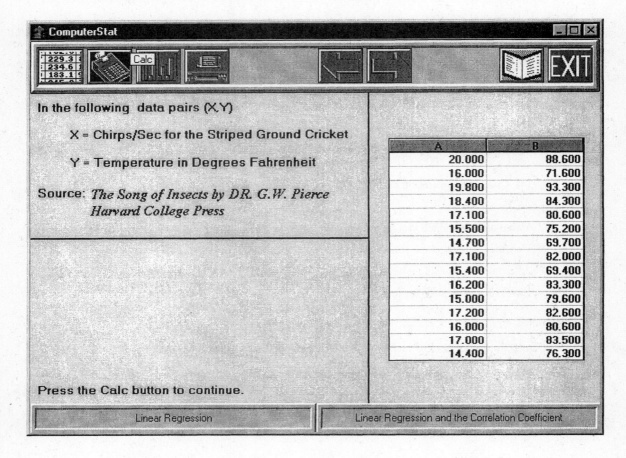

Click on the calculate button.

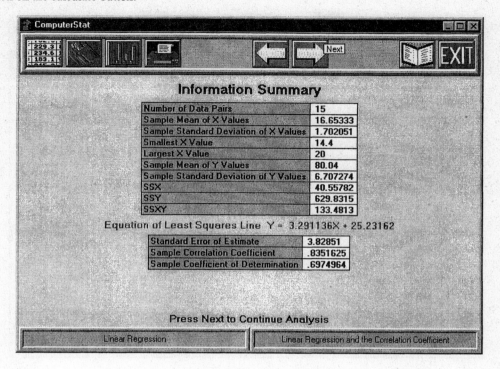

Notice that the value for the standard error of estimate $S_e \approx 3.829$. Clicking on the Next button gives you a choice of graphing the scatter diagram and least-squares line, predicting Y from an X value, or testing the correlation coefficient R.

Select the Prediction option. For X = 18 and a 90% confidence interval for the prediction, we see

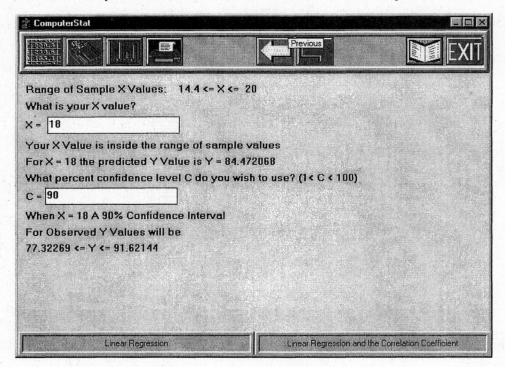

Clicking Print will print your results. Click on the "previous" button and choose the option to test the correlation coefficient.

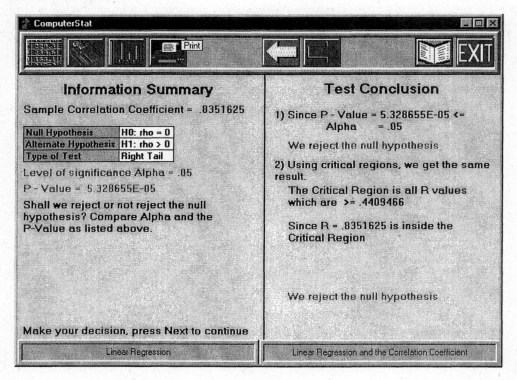

LAB ACTIVITIES FOR LINEAR REGRESSION AND CORRELATION

1. Select Class Demonstration #1: List Price vs. Best Price for a New GMC Pickup Truck.

In the Following Data Pairs (X, Y),
X = List Price (in $1000) for a GMC Pickup Truck
Y = Best Price (in $1000) for a GMC Pickup Truck

(12.400, 11.200)	(14.300, 12.500)	(14.500, 12.700)
(14.900, 13.100)	(16.100, 14.100)	(16.900, 14.800)
(16.500, 14.400)	(15.400, 13.400)	(17.000, 14.900)
(17.900, 15.600)	(18.800, 16.400)	(20.300, 17.700)
(22.400, 19.600)	(19.400, 16.900)	(15.500, 14.000)
(16.700, 14.600)	(17.300, 15.100)	(18.400, 16.100)
(19.200, 16.800)	(17.400, 15.200)	(19.500, 17.000)
(19.700, 17.200)	(21.200, 18.600)	

(a) Look at the scatter diagram. Do you think linear regression is appropriate for this data?

(b) Look at the information summary sheet and write down the equation of the least-squares line, the value of the correlation coefficient, the value of the coefficient of determination, and the value of the standard error of estimate.

(c) If the list price of a pickup truck is $20,000, what is the predicted best price? (Note: Be sure to change 20,000 to 20 thousand, since the data have been entered in thousands.) Find a 90% confidence interval for the prediction. Would you be surprised if the best price you could get were $19,500?

(d) Test the correlation coefficient. At the 5% level of significance, is there evidence of a positive correlation between list and best price?

2. Merchandise loss due to shoplifting, damage, and other causes is called shrinkage. Shrinkage is a major concern to retailers. The managers of H.R. Merchandise think there is a relationship between shrinkage and number of clerks on duty. To explore this relationship, a random sample of 7 weeks was selected. During each week the staffing level of sale clerks was kept constant and the dollar value of the shrinkage was recorded. The results follow.

Number of clerks, X	10	12	11	15	9	13	8
Shrinkage, Y (in $hundreds)	19	15	20	9	25	12	31

(a) Select the option to enter your own data. Enter the data. Look at the scatter diagram. Does a linear regression model seem appropriate for this model?

(b) Look at the information summary sheet and write down the equation of the least-squares line, the value of the correlation coefficient, the value of the coefficient of determination, and the value of the standard error of estimate.

(c) Does it make sense to extrapolate beyond the data? Why or why not? What does the model predict for shrinkage if there is a staff of 40 clerks? What about for a staff of 1 clerk? Can you get confidence intervals for these predictions?

(d) What does the model predict for a staffing level of 12 clerks? Find a 90% confidence interval for shrinkage.

(e) Test the correlation coefficient. Does the sample provide evidence of a negative correlation between staffing level and shrinkage?

Descriptions of Data Sets
on the
HM StatPass CD-ROM

I.   General

There are over 100 data sets saved in Excel, Minitab Portable, and TI-83 Plus/ASCII formats to accompany *Understanding Basic Statistics,* 3rd edition. These files can be found on the HM StatPass CD-ROM packaged with the student textbook. You may also visit the Brase/Brase statistics site at http://math.college.hmco.com/students to view these data sets online. The data sets are organized by category.

A. The following are provided for each data set:
1. The category
2. A brief description of the data and variables with a reference when appropriate
3. File names for Excel, Minitab, and TI-83Plus/ASCII formats

B. The categories are
1. **Single variable large sample** ($n \geq 30$)
    File name prefix **Svls** followed by the data set number
    There are 30 data sets...   page 109
2. **Single variable small sample** ($n < 30$)
    File name prefix **Svss** followed by the data set number
    There are 11 data sets...   page 120
3. **Time series data** for control chart about the mean or for P-Charts
    File name prefix **Tscc** followed by the data set number
    There are 10 data sets...   page 123
4. **Two variable independent samples** (large and small sample)
    File name prefix **Tvis** followed by the data set number
    There are 10 data sets...   page 127
5. **Two variable dependent samples** appropriate for *t*-tests
    File name prefix **Tvds** followed by the data set number
    There are 10 data sets...   page 132
6. **Simple linear regression**
    File name prefix **Slr** followed by the data set number
    There are 12 data sets...   page 137
7. **Multiple linear regression**
    File name prefix **Mlr** followed by the data set number
    There are 10 data sets...   page 144
8. **One-way ANOVA**
    File name prefix **Owan** followed by the data set number
    There are 5 data sets...   page 158
9. **Two-way ANOVA**
    File name prefix **Twan** followed by the data set number
    There are 5 data sets...   page 163

C. The formats are
1. Excel files in subdirectory Excel_7e. These files have suffix .xls
2. Minitab portable files in subdirectory Minitab_7e. These files have suffix .mtp
3. TI-83Plus/ASCII files in subdirectory TI83_7e. These files have suffix .txt
4. Some of the data sets are also built-in to ComputerStat, which is included on the HM StatPass CD-ROM.

II. Suggestions for using the data sets

1. Single variable large sample (file name prefix Svls)
   These data sets are appropriate for
   > Graphs: Histograms, box plots
   > Descriptive statistics: Mean, median, mode, variance, standard deviation, coefficient of variation, 5 number summary
   > Inferential statistics: Confidence intervals for the population mean, hypothesis tests of a single mean

2. Single variable small sample (file name prefix Svss)
   > Graphs: Histograms, box plots,
   > Descriptive statistics: Mean, median, mode, variance, standard deviation, coefficient of variation, 5 number summary
   > Inferential statistics: Confidence intervals for the population mean, hypothesis tests of a single mean

3. Time series data (file name prefix Tsccc)
   > Graphs: Time plots, control charts about the mean utilizing individual data for the data sets so designated, P charts for the data sets so designated

4. Two independent data sets (file name prefix Tvis)
   > Graphs: Histograms, box plots for each data set
   > Descriptive statistics: Mean, median, mode, variance, standard deviation, coefficient of variation, 5-number summary for each data set
   > Inferential statistics: Confidence intervals for the difference of means, hypothesis tests for the difference of means

5. Paired data, dependent samples (file name prefix Tvds)
   > Descriptive statistics: Mean, median, mode, variance, standard deviation, coefficient of variation, 5 number summary for the difference of the paired data values.
   > Inferential statistics: Hypothesis tests for the difference of means (paired data)

6. Data pairs for simple linear regression (file name prefix Slr)
   > Graphs: Scatter plots, for individual variables histograms and box plots
   > Descriptive statistics:
   > (a) Mean, median, mode, variance, standard deviation, coefficient of variation, 5 number summary for individual variables.
   > (b) Least squares line, sample correlation coefficient, sample coefficient of determination
   > Inferential statistics: Testing ρ, confidence intervals for β, testing β

7. Data for multiple linear regression (file name prefix Mlr)
   > Graphs:
   > Descriptive statistics: Histograms, box plots for individual variables
   > (a) Mean, median, mode, variance, standard deviation, coefficient of variation, 5 number summary for individual variables.
   > (b) Least squares line, sample coefficient of determination
   > Inferential statistics: confidence intervals for coefficients, testing coefficients

8. Data for one-way ANOVA (file name prefix Owan)
   - Graphs: Histograms, box plots for individual samples
   - Descriptive statistics: Mean, median, mode, variance, standard deviation, coefficient of variation, 5 number summary for individual samples.
   - Inferential statistics: One-way ANOVA

9. Data for two-way ANOVA (file name prefix Twan)
   - Graphs: Histograms, box plots for individual samples
   - Descriptive statistics: Mean, median, mode, variance, standard deviation, coefficient of variation, 5 number summary for data in individual cells.
   - Inferential statistics: Two-way ANOVA

Single Variable Large Sample ($n \geq 30$)
File name prefix: Svls followed by the number of the data file

01. **Disney Stock Volume (Single Variable Large Sample $n \geq 30$)**
    The following data represents the number of shares of Disney stock (in hundreds of shares) sold for a random sample of 60 trading days
    Reference: *The Denver Post*, Business section

12584	9441	18960	21480	10766	13059	8589	4965
4803	7240	10906	8561	6389	14372	18149	6309
13051	12754	10860	9574	19110	29585	21122	14522
17330	18119	10902	29158	16065	10376	10999	17950
15418	12618	16561	8022	9567	9045	8172	13708
11259	10518	9301	5197	11259	10518	9301	5197
6758	7304	7628	14265	13054	15336	14682	27804
16022	24009	32613	19111				

    File names   Excel: Svls01.xls
                        Minitab: Svls01.mtp
                        TI-83Plus/ASCII: Svls01.txt

02. **Weights of Pro Football Players (Single Variable Large Sample $n \geq 30$)**
    The following data represents weights in pounds of 50 randomly selected pro football linebackers.
    Reference: The Sports Encyclopedia Pro Football

225	230	235	238	232	227	244	222
250	226	242	253	251	225	229	247
239	223	233	222	243	237	230	240
255	230	245	240	235	252	245	231
235	234	248	242	238	240	240	240
235	244	247	250	236	246	243	255
241	245						

    File names   Excel: Svls02.xls
                        Minitab: Svls02.mtp
                        TI-83Plus/ASCII: Svls02.txt

03. Heights of Pro Basketball Players (Single Variable Large Sample $n \geq 30$)
The following data represents heights in feet of 65 randomly selected pro basketball players.
Reference: All-Time Player Directory, The Official NBA Encyclopedia

6.50	6.25	6.33	6.50	6.42	6.67	6.83	6.82
6.17	7.00	5.67	6.50	6.75	6.54	6.42	6.58
6.00	6.75	7.00	6.58	6.29	7.00	6.92	6.42
5.92	6.08	7.00	6.17	6.92	7.00	5.92	6.42
6.00	6.25	6.75	6.17	6.75	6.58	6.58	6.46
5.92	6.58	6.13	6.50	6.58	6.63	6.75	6.25
6.67	6.17	6.17	6.25	6.00	6.75	6.17	6.83
6.00	6.42	6.92	6.50	6.33	6.92	6.67	6.33
6.08							

File names   Excel: Svls03.xls
                      Minitab: Svls03.mtp
                      TI-83Plus/ASCII: Svls03.txt

04. Miles per Gallon Gasoline Consumption (Single Variable Large Sample $n \geq 30$)
The following data represents miles per gallon gasoline consumption (highway) for a random sample of 55 makes and models of passenger cars.
Reference: Environmental Protection Agency

30	27	22	25	24	25	24	15
35	35	33	52	49	10	27	18
20	23	24	25	30	24	24	24
18	20	25	27	24	32	29	27
24	27	26	25	24	28	33	30
13	13	21	28	37	35	32	33
29	31	28	28	25	29	31	

File names   Excel: Svls04.xls
                      Minitab: Svls04.mtp
                      TI-83Plus/ASCII: Svls04.txt

05. Fasting Glucose Blood Tests (Single Variable Large Sample $n \geq 30$)
The following data represents glucose blood level (mg/100ml) after a 12-hour fast for a random sample of 70 women.
Reference: American J. Clin. Nutr. Vol. 19, 345-351

45	66	83	71	76	64	59	59
76	82	80	81	85	77	82	90
87	72	79	69	83	71	87	69
81	76	96	83	67	94	101	94
89	94	73	99	93	85	83	80
78	80	85	83	84	74	81	70
65	89	70	80	84	77	65	46
80	70	75	45	101	71	109	73
73	80	72	81	63	74		

File names   Excel: Svls05.xls
                      Minitab: Svls05.mtp
                      TI-83Plus/ASCII: Svls05.txt

06. Number of Children in Rural Canadian Families (Single Variable Large Sample $n \geq 30$)
The following data represents the number of children in a random sample of 50 rural Canadian families.
Reference: American Journal Of Sociology Vol. 53, 470-480

11	13	4	14	10	2	5	0
0	3	9	2	5	2	3	3
3	4	7	1	9	4	3	3
2	6	0	2	6	5	9	5
4	3	2	5	2	2	3	5
14	7	6	6	2	5	3	4
6	1						

File names	Excel: Svls06.xls
	Minitab: Svls06.mtp
	TI-83Plus/ASCII: Svls06.txt

07. Children as a % of Population (Single Variable Large Sample $n \geq 30$)
The following data represent percentage of children in the population for a random sample of 72 Denver neighborhoods.
Reference: The Piton Foundation, Denver, Colorado

30.2	18.6	13.6	36.9	32.8	19.4	12.3	39.7	22.2	31.2
36.4	37.7	38.8	28.1	18.3	22.4	26.5	20.4	37.6	23.8
22.1	53.2	6.8	20.7	31.7	10.4	21.3	19.6	41.5	29.8
14.7	12.3	17.0	16.7	20.7	34.8	7.5	19.0	27.2	16.3
24.3	39.8	31.1	34.3	15.9	24.2	20.3	31.2	30.0	33.1
29.1	39.0	36.0	31.8	32.9	26.5	4.9	19.5	21.0	24.2
12.1	38.3	39.3	20.2	24.0	28.6	27.1	30.0	60.8	39.2
21.6	20.3								

File names	Excel: Svls07.xls
	Minitab: Svls07.mtp
	TI-83Plus/ASCII: Svls07.txt

08. Percentage Change in Household Income (Single Variable Large Sample $n \geq 30$)
The following data represent the percentage change in household income over a five-year period for a random sample of $n = 78$ Denver neighborhoods.
Reference: The Piton Foundation, Denver, Colorado

27.2	25.2	25.7	80.9	26.9	20.2	25.4	26.9	26.4	26.3
27.5	38.2	20.9	31.3	23.5	26.0	35.8	30.9	15.5	24.8
29.4	11.7	32.6	32.2	27.6	27.5	28.7	28.0	15.6	20.0
21.8	18.4	27.3	13.4	14.7	21.6	26.8	20.9	32.7	29.3
21.4	29.0	7.2	25.7	25.5	39.8	26.6	24.2	33.5	16.0
29.4	26.8	32.0	24.7	24.2	29.8	25.8	18.2	26.0	26.2
21.7	27.0	23.7	28.0	11.2	26.2	21.6	23.7	28.3	34.1
40.8	16.0	50.5	54.1	3.3	23.5	10.1	14.8		

File names	Excel: Svls08.xls
	Minitab: Svls08.mtp
	TI-83Plus/ASCII: Svls08.txt

09. Crime Rate per 1,000 Population (Single Variable Large Sample $n \geq 30$)

The following data represent the crime rate per 1,000 population for a random sample of 70 Denver neighborhoods.

Reference: The Piton Foundation, Denver, Colorado

84.9	45.1	132.1	104.7	258.0	36.3	26.2	207.7
58.5	65.3	42.5	53.2	172.6	69.2	179.9	65.1
32.0	38.3	185.9	42.4	63.0	86.4	160.4	26.9
154.2	111.0	139.9	68.2	127.0	54.0	42.1	105.2
77.1	278.0	73.0	32.1	92.7	704.1	781.8	52.2
65.0	38.6	22.5	157.3	63.1	289.1	52.7	108.7
66.3	69.9	108.7	96.9	27.1	105.1	56.2	80.1
59.6	77.5	68.9	35.2	65.4	123.2	130.8	70.7
25.1	62.6	68.6	334.5	44.6	87.1		

File names   Excel: Svls09.xls
Minitab: Svls09.mtp
TI-83Plus/ASCII: Svls09.txt

10. Percentage Change in Population (Single Variable Large Sample $n \geq 30$)

The following data represent the percentage change in population over a nine-year period for a random sample of 64 Denver neighborhoods.

Reference: The Piton Foundation, Denver, Colorado

6.2	5.4	8.5	1.2	5.6	28.9	6.3	10.5	-1.5	17.3
21.6	-2.0	-1.0	3.3	2.8	3.3	28.5	-0.7	8.1	32.6
68.6	56.0	19.8	7.0	38.3	41.2	4.9	7.8	7.8	97.8
5.5	21.6	32.5	-0.5	2.8	4.9	8.7	-1.3	4.0	32.2
2.0	6.4	7.1	8.8	3.0	5.1	-1.9	-2.6	1.6	7.4
10.8	4.8	1.4	19.2	2.7	71.4	2.5	6.2	2.3	10.2
1.9	2.3	-3.3	2.6						

File names   Excel: Svls10.xls
Minitab: Svls10.mtp
TI-83Plus/ASCII: Svls10.txt

11. Thickness of the Ozone Column (Single Variable Large Sample $n \geq 30$)

The following data represent the January mean thickness of the ozone column above Arosa, Switzerland (Dobson units: one milli-centimeter ozone at standard temperature and pressure). The data is from a random sample of years from 1926 on.

Reference: Laboratorium fuer Atmosphaerensphysik, Switzerland

324	332	362	383	335	349	354	319	360	329
400	341	315	368	361	336	349	347	338	332
341	352	342	361	318	337	300	352	340	371
327	357	320	377	338	361	301	331	334	387
336	378	369	332	344					

File names   Excel: Svls11.xls
Minitab: Svls11.mtp
TI-83Plus/ASCII: Svls11.txt

12. **Sun Spots (Single Variable Large Sample $n \geq 30$)**
The following data represent the January mean number of sunspots. The data is taken from a random sample of Januarys from 1749 to 1983.
Reference: Waldmeir, M, *Sun Spot Activity*, International Astronomical Union Bulletin

12.5	14.1	37.6	48.3	67.3	70.0	43.8	56.5	59.7	24.0
12.0	27.4	53.5	73.9	104.0	54.6	4.4	177.3	70.1	54.0
28.0	13.0	6.5	134.7	114.0	72.7	81.2	24.1	20.4	13.3
9.4	25.7	47.8	50.0	45.3	61.0	39.0	12.0	7.2	11.3
22.2	26.3	34.9	21.5	12.8	17.7	34.6	43.0	52.2	47.5
30.9	11.3	4.9	88.6	188.0	35.6	50.5	12.4	3.7	18.5
115.5	108.5	119.1	101.6	59.9	40.7	26.5	23.1	73.6	165.0
202.5	217.4	57.9	38.7	15.3	8.1	16.4	84.3	51.9	58.0
74.7	96.0	48.1	51.1	31.5	11.8	4.5	78.1	81.6	68.9

File names   Excel: Svls12.xls
                    Minitab: Svls12.mtp
                    TI-83Plus/ASCII: Svls12.txt

13. **Motion of Stars (Single Variable Large Sample $n \geq 30$)**
The following data represent the angular motions of stars across the sky due to the stars own velocity. A random sample of stars from the M92 global cluster was used. Units are arc seconds per century.
Reference: Cudworth, K.M., Astronomical Journal, Vol. 81, p 975-982

0.042	0.048	0.019	0.025	0.028	0.041	0.030	0.051	0.026
0.040	0.018	0.022	0.048	0.045	0.019	0.028	0.029	0.018
0.033	0.035	0.019	0.046	0.021	0.026	0.026	0.033	0.046
0.023	0.036	0.024	0.014	0.012	0.037	0.034	0.032	0.035
0.015	0.027	0.017	0.035	0.021	0.016	0.036	0.029	0.031
0.016	0.024	0.015	0.019	0.037	0.016	0.024	0.029	0.025
0.022	0.028	0.023	0.021	0.020	0.020	0.016	0.016	0.016
0.040	0.029	0.025	0.025	0.042	0.022	0.037	0.024	0.046
0.016	0.024	0.028	0.027	0.060	0.045	0.037	0.027	0.028
0.022	0.048	0.053						

File names   Excel: Svls13.xls
                    Minitab: Svls13.mtp
                    TI-83Plus/ASCII: Svls13.txt

14. **Arsenic and Ground Water (Single Variable Large Sample $n \geq 30$)**
    The following data represent (naturally occurring) concentration of arsenic in ground water for a random sample of 102 Northwest Texas wells. Units are parts per billion.
    Reference: Nichols, C.E. and Kane, V.E., Union Carbide Technical Report K/UR-1

7.6	10.4	13.5	4.0	19.9	16.0	12.0	12.2	11.4	12.7
3.0	10.3	21.4	19.4	9.0	6.5	10.1	8.7	9.7	6.4
9.7	63.0	15.5	10.7	18.2	7.5	6.1	6.7	6.9	0.8
73.5	12.0	28.0	12.6	9.4	6.2	15.3	7.3	10.7	15.9
5.8	1.0	8.6	1.3	13.7	2.8	2.4	1.4	2.9	13.1
15.3	9.2	11.7	4.5	1.0	1.2	0.8	1.0	2.4	4.4
2.2	2.9	3.6	2.5	1.8	5.9	2.8	1.7	4.6	5.4
3.0	3.1	1.3	2.6	1.4	2.3	1.0	5.4	1.8	2.6
3.4	1.4	10.7	18.2	7.7	6.5	12.2	10.1	6.4	10.7
6.1	0.8	12.0	28.1	9.4	6.2	7.3	9.7	62.1	15.5
6.4	9.5								

    File names

    Excel: Svls14.xls
    Minitab: Svls14.mtp
    TI-83Plus/ASCII: Svls14.txt

15. **Uranium In Ground Water (Single Variable Large Sample $n \geq 30$)**
    The following data represent (naturally occurring) concentrations of uranium in ground water for a random sample of 100 Northwest Texas wells. Units are parts per billion.
    Reference: Nichols, C.E. and Kane, V.E., Union Carbide Technical Report K/UR-1

8.0	13.7	4.9	3.1	78.0	9.7	6.9	21.7	26.8
56.2	25.3	4.4	29.8	22.3	9.5	13.5	47.8	29.8
13.4	21.0	26.7	52.5	6.5	15.8	21.2	13.2	12.3
5.7	11.1	16.1	11.4	18.0	15.5	35.3	9.5	2.1
10.4	5.3	11.2	0.9	7.8	6.7	21.9	20.3	16.7
2.9	124.2	58.3	83.4	8.9	18.1	11.9	6.7	9.8
15.1	70.4	21.3	58.2	25.0	5.5	14.0	6.0	11.9
15.3	7.0	13.6	16.4	35.9	19.4	19.8	6.3	2.3
1.9	6.0	1.5	4.1	34.0	17.6	18.6	8.0	7.9
56.9	53.7	8.3	33.5	38.2	2.8	4.2	18.7	12.7
3.8	8.8	2.3	7.2	9.8	7.7	27.4	7.9	11.1
24.7								

    File names

    Excel: Svls15.xls
    Minitab: Svls15.mtp
    TI-83Plus/ASCII: Svls15.txt

16. **Ground Water pH (Single Variable Large Sample $n \geq 30$)**
    A pH less than 7 is acidic, and a pH above 7 is alkaline. The following data represent pH levels in ground
    water for a random sample of 102 Northwest Texas wells.
    Reference: Nichols, C.E. and Kane, V.E., Union Carbide Technical Report K/UR-1

7.6	7.7	7.4	7.7	7.1	8.2	7.4	7.5	7.2	7.4
7.2	7.6	7.4	7.8	8.1	7.5	7.1	8.1	7.3	8.2
7.6	7.0	7.3	7.4	7.8	8.1	7.3	8.0	7.2	8.5
7.1	8.2	8.1	7.9	7.2	7.1	7.0	7.5	7.2	7.3
8.6	7.7	7.5	7.8	7.6	7.1	7.8	7.3	8.4	7.5
7.1	7.4	7.2	7.4	7.3	7.7	7.0	7.3	7.6	7.2
8.1	8.2	7.4	7.6	7.3	7.1	7.0	7.0	7.4	7.2
8.2	8.1	7.9	8.1	8.2	7.7	7.5	7.3	7.9	8.8
7.1	7.5	7.9	7.5	7.6	7.7	8.2	8.7	7.9	7.0
8.8	7.1	7.2	7.3	7.6	7.1	7.0	7.0	7.3	7.2
7.8	7.6								

    File names   Excel: Svls16.xls
                          Minitab: Svls16.mtp
                          TI-83Plus/ASCII: Svls16.txt

17. **Static Fatigue 90% Stress Level (Single Variable Large Sample $n \geq 30$)**
    Kevlar Epoxy is a material used on the NASA space shuttle. Strands of this epoxy were tested at 90%
    breaking strength. The following data represent time to failure in hours at the 90% stress level for a
    random sample of 50 epoxy strands.
    Reference: R.E. Barlow University of California, Berkeley

0.54	1.80	1.52	2.05	1.03	1.18	0.80	1.33	1.29	1.11
3.34	1.54	0.08	0.12	0.60	0.72	0.92	1.05	1.43	3.03
1.81	2.17	0.63	0.56	0.03	0.09	0.18	0.34	1.51	1.45
1.52	0.19	1.55	0.02	0.07	0.65	0.40	0.24	1.51	1.45
1.60	1.80	4.69	0.08	7.89	1.58	1.64	0.03	0.23	0.72

    File names   Excel: Svls17.xls
                          Minitab: Svls17.mtp
                          TI-83Plus/ASCII: Svls17.txt

18. **Static Fatigue 80% Stress Level (Single Variable Large Sample $n \geq 30$)**
    Kevlar Epoxy is a material used on the NASA space shuttle. Strands of this epoxy were tested at 80%
    breaking strength. The following data represent time to failure in hours at the 80% stress level for a
    random sample of 54 epoxy strands.
    Reference: R.E. Barlow University of California, Berkeley

152.2	166.9	183.8	8.5	1.8	118.0	125.4	132.8	10.6
29.6	50.1	202.6	177.7	160.0	87.1	112.6	122.3	124.4
131.6	140.9	7.5	41.9	59.7	80.5	83.5	149.2	137.0
301.1	329.8	461.5	739.7	304.3	894.7	220.2	251.0	269.2
130.4	77.8	64.4	381.3	329.8	451.3	346.2	663.0	49.1
31.7	116.8	140.2	334.1	285.9	59.7	44.1	351.2	93.2

    File names   Excel: Svls18.xls
                          Minitab: Svls18.mtp
                          TI-83Plus/ASCII: Svls18.txt

19. **Tumor Recurrence (Single Variable Large Sample $n \geq 30$)**
Certain kinds of tumors tend to recur. The following data represents the length of time in months for a tumor to recur after chemotherapy (sample size, 42).
Reference: Byar, D.P, *Urology* Vol. 10, p 556-561

19	18	17	1	21	22	54	46	25	49
50	1	59	39	43	39	5	9	38	18
14	45	54	59	46	50	29	12	19	36
38	40	43	41	10	50	41	25	19	39
27	20								

File names	Excel: Svls19.xls
	Minitab: Svls19.mtp
	TI-83Plus/ASCII: Svls19.txt

20. **Weight of Harvest (Single Variable Large Sample $n \geq 30$)**
The following data represent the weights in kilograms of maize harvest from a random sample of 72 experimental plots on the island of St Vincent (Caribbean).
Reference: Springer, B.G.F. *Proceedings, Caribbean Food Corps. Soc.* Vol. 10 p 147-152

24.0	27.1	26.5	13.5	19.0	26.1	23.8	22.5	20.0
23.1	23.8	24.1	21.4	26.7	22.5	22.8	25.2	20.9
23.1	24.9	26.4	12.2	21.8	19.3	18.2	14.4	22.4
16.0	17.2	20.3	23.8	24.5	13.7	11.1	20.5	19.1
20.2	24.1	10.5	13.7	16.0	7.8	12.2	12.5	14.0
22.0	16.5	23.8	13.1	11.5	9.5	22.8	21.1	22.0
11.8	16.1	10.0	9.1	15.2	14.5	10.2	11.7	14.6
15.5	23.7	25.1	29.5	24.5	23.2	25.5	19.8	17.8

File names	Excel: Svls20.xls
	Minitab: Svls20.mtp
	TI-83Plus/ASCII: Svls20.txt

21. **Apple Trees (Single Variable Large Sample $n \geq 30$)**
The following data represent trunk girth (mm) of a random sample of 60 four-year-old apple trees at East Malling Research Station (England)
Reference: S.C. Pearce, University of Kent at Canterbury

108	99	106	102	115	120	120	117	122	142
106	111	119	109	125	108	116	105	117	123
103	114	101	99	112	120	108	91	115	109
114	105	99	122	106	113	114	75	96	124
91	102	108	110	83	90	69	117	84	142
122	113	105	112	117	122	129	100	138	117

File names	Excel: Svls21.xls
	Minitab: Svls21.mtp
	TI-83Plus/ASCII: Svls21.txt

22. **Black Mesa Archaeology (Single Variable Large Sample $n \geq 30$)**
The following data represent rim diameters (cm) of a random sample of 40 bowls found at Black Mesa archaeological site. The diameters are estimated from broken pot shards.
Reference: Michelle Hegmon, Crow Canyon Archaeological Center, Cortez, Colorado

```
17.2  15.1  13.8  18.3  17.5  11.1   7.3  23.1  21.5  19.7
17.6  15.9  16.3  25.7  27.2  33.0  10.9  23.8  24.7  18.6
16.9  18.8  19.2  14.6   8.2   9.7  11.8  13.3  14.7  15.8
17.4  17.1  21.3  15.2  16.8  17.0  17.9  18.3  14.9  17.7
```

File names   Excel: Svls22.xls
              Minitab: Svls22.mtp
              TI-83Plus/ASCII: Svls22.txt

23. **Wind Mountain Archaeology (Single Variable Large Sample $n \geq 30$)**
The following data represent depth (cm) for a random sample of 73 significant archaeological artifacts at the Wind Mountain excavation site.
Reference: Woosley, A. and McIntyre, A. *Mimbres Mogolion Archaology*, University New Mexico press.

```
85   45   75   60   90   90  115   30   55   58
78  120   80   65   65  140   65   50   30  125
75  137   80  120   15   45   70   65   50   45
95   70   70   28   40  125  105   75   80   70
90   68   73   75   55   70   95   65  200   75
15   90   46   33  100   65   60   55   85   50
10   68   99  145   45   75   45   95   85   65
65   52   82
```

File names   Excel: Svls23.xls
              Minitab: Svls23.mtp
              TI-83Plus/ASCII: Svls23.txt

24. **Arrow Heads (Single Variable Large Sample $n \geq 30$)**
The following data represent length (cm) of a random sample of 61 projectile points found at the Wind Mountain Archaeological site.
Reference: Woosley, A. and McIntyre, A. *Mimbres Mogolion Archaology*, University New Mexico press.

```
3.1  4.1  1.8  2.1  2.2  1.3  1.7  3.0  3.7  2.3
2.6  2.2  2.8  3.0  3.2  3.3  2.4  2.8  2.8  2.9
2.9  2.2  2.4  2.1  3.4  3.1  1.6  3.1  3.5  2.3
3.1  2.7  2.1  2.0  4.8  1.9  3.9  2.0  5.2  2.2
2.6  1.9  4.0  3.0  3.4  4.2  2.4  3.5  3.1  3.7
3.7  2.9  2.6  3.6  3.9  3.5  1.9  4.0  4.0  4.6
1.9
```

File names   Excel: Svls24.xls
              Minitab: Svls24.mtp
              TI-83Plus/ASCII: Svls24.txt

25. **Anasazi Indian Bracelets (Single Variable Large Sample $n \geq 30$)**
The following data represent the diameter (cm) of shell bracelets and rings found at the Wind Mountain archaeological site.
Reference: Woosley, A. and McIntyre, A. *Mimbres Mogolion Archaology*, University New Mexico press.

```
5.0   5.0   8.0   6.1   6.0   5.1   5.9   6.8   4.3   5.5
7.2   7.0   5.0   5.6   5.3   7.0   3.4   8.2   4.3   5.2
1.5   6.1   4.0   6.0   5.5   5.2   5.2   5.2   5.5   7.2
6.0   6.2   5.2   5.0   4.0   5.7   5.1   6.1   5.7   7.3
7.3   6.7   4.2   4.0   6.0   7.1   7.3   5.5   5.8   8.9
7.5   8.3   6.8   4.9   4.0   6.2   7.7   5.0   5.2   6.8
6.1   7.2   4.4   4.0   5.0   6.0   6.2   7.2   5.8   6.8
7.7   4.7   5.3
```

File names	Excel: Svls25.xls
	Minitab: Svls25.mtp
	TI-83Plus/ASCII: Svls25.txt

26. **Pizza Franchise Fees (Single Variable Large Sample $n \geq 30$)**
The following data represent annual franchise fees (in thousands of dollars) for a random sample of 36 pizza franchises.
Reference: *Business Opportunities Handbook*

```
25.0   15.5    7.5   19.9   18.5   25.5   15.0    5.5   15.2   15.0
14.9   18.5   14.5   29.0   22.5   10.0   25.0   35.5   22.1   89.0
17.5   33.3   17.5   12.0   15.5   25.5   12.5   17.5   12.5   35.0
30.0   21.0   35.5   10.5    5.5   20.0
```

File names	Excel: Svls26.xls
	Minitab: Svls26.mtp
	TI-83Plus/ASCII: Svls26.txt

27. **Pizza Franchise Start-up Requirement (Single Variable Large Sample $n \geq 30$)**
The following data represent annual start-up cost (in thousands of dollars) for a random sample of 36 pizza franchises.
Reference: *Business Opportunities Handbook*

```
40    25    50   129   250   128   110   142    25    90
75   100   500   214   275    50   128   250    50    75
30    40   185    50   175   125   200   150   150   120
95    30   400   149   235   100
```

File names	Excel: Svls27.xls
	Minitab: Svls27.mtp
	TI-83Plus/ASCII: Svls27.txt

28. College Degrees (Single Variable Large Sample $n \geq 30$)
The following data represent percentages of the adult population with college degrees. The sample is from a random sample of 68 Midwest counties.
Reference: *County and City Data Book* 12th edition, U.S. Department of Commerce

```
 9.9   9.8   6.8   8.9  11.2  15.5   9.8  16.8   9.9  11.6
 9.2   8.4  11.3  11.5  15.2  10.8  16.3  17.0  12.8  11.0
 6.0  16.0  12.1   9.8   9.4   9.9  10.5  11.8  10.3  11.1
12.5   7.8  10.7   9.6  11.6   8.8  12.3  12.2  12.4  10.0
10.0  18.1   8.8  17.3  11.3  14.5  11.0  12.3   9.1  12.7
 5.6  11.7  16.9  13.7  12.5   9.0  12.7  11.3  19.5  30.7
 9.4   9.8  15.1  12.8  12.9  17.5  12.3   8.2
```

File names   Excel: Svls28.xls
                           Minitab: Svls28.mtp
                           TI-83Plus/ASCII: Svls28.txt

29. Poverty Level (Single Variable Large Sample $n \geq 30$)
The following data represent percentages of all persons below the poverty level. The sample is from a random collection of 80 cities in the Western U.S.
Reference: *County and City Data Book* 12th edition, U.S. Department of Commerce

```
12.1  27.3  20.9  14.9   4.4  21.8   7.1  16.4  13.1
 9.4   9.8  15.7  29.9   8.8  32.7   5.1   9.0  16.8
21.6   4.2  11.1  14.1  30.6  15.4  20.7  37.3   7.7
19.4  18.5  19.5   8.0   7.0  20.2   6.3  12.9  13.3
30.0   4.9  14.4  14.1  22.6  18.9  16.8  11.5  19.2
21.0  11.4   7.8   6.0  37.3  44.5  37.1  28.7   9.0
17.9  16.0  20.2  11.5  10.5  17.0   3.4   3.3  15.6
16.6  29.6  14.9  23.9  13.6   7.8  14.5  19.6  31.5
28.1  19.2   4.9  12.7  15.1   9.6  23.8  10.1
```

File names   Excel: Svls29.xls
                           Minitab: Svls29.mtp
                           TI-83Plus/ASCII: Svls29.txt

30. Working at Home (Single Variable Large Sample $n \geq 30$)
The following data represent percentages of adults whose primary employment involves working at home. The data is from a random sample of 50 California cities.
Reference: *County and City Data Book* 12th edition, U.S. Department of Commerce

```
4.3   5.1          3.1   8.7   4.0   5.2  11.8   3.4   8.5   3.0
4.3   6.0          3.7   3.7   4.0   3.3   2.8   2.8   2.6   4.4
7.0   8.0          3.7   3.3   3.7   4.9   3.0   4.2   5.4   6.6
2.4   2.5          3.5   3.3   5.5   9.6   2.7   5.0   4.8   4.1
3.8   4.8  14.3   9.2   3.8   3.6   6.5   2.6   3.5   8.6
```

File names   Excel: Svls30.xls
                           Minitab: Svls30.mtp
                           TI-83Plus/ASCII: Svls30.txt

Single Variable Small Sample ($n < 30$)
File name prefix: SVSS followed by the number of the data file

01. Number of Pups in Wolf Den (Single Variable Small Sample $n < 30$)
The following data represent the number of wolf pups per den from a random sample of 16 wolf dens.
Reference: *The Wolf in the Southwest: The Making of an Endangered Species*, Brown, D.E., University of Arizona Press

```
5   8   7   5   3   4   3   9
5   8   5   6   5   6   4   7
```

File names   Excel: Svss01.xls
                    Minitab: Svss01.mtp
                    TI-83plus/ASCII: Svss01.txt

02. Glucose Blood Level (Single Variable Small Sample $n < 30$)
The following data represent glucose blood level (mg/100ml) after a 12-hour fast for a random sample of 6 tests given to an individual adult female.
Reference: *American J. Clin. Nutr. Vol. 19*, p345-351

```
83   83   86   86   78   88
```

File names   Excel: Svss02.xls
                    Minitab: Svss02.mtp
                    TI-83plus/ASCII: Svss02.txt

03. Length of Remission (Single Variable Small Sample $n < 30$)
The drug 6-mP (6-mercaptopurine) is used to treat leukemia. The following data represent the length of remission in weeks for a random sample of 21 patients using 6-mP.
Reference: E.A. Gehan, University of Texas Cancer Center

```
10    7   32   23   22    6   16   34   32   25
11   20   19    6   17   35    6   13    9    6
10
```

File names   Excel: Svss03.xls
                    Minitab: Svss03.mtp
                    TI-83plus/ASCII: Svss03.txt

04. Entry Level Jobs (Single Variable Small Sample $n < 30$)
The following data represent percentage of entry-level jobs in a random sample of 16 Denver neighborhoods.
Reference: The Piton Foundation, Denver, Colorado

```
8.9   22.6   18.5    9.2    8.2   24.3   15.3   3.7
9.2   14.9    4.7   11.6   16.5   11.6    9.7   8.0
```

File names   Excel: Svss04.xls
                    Minitab: Svss04.mtp
                    TI-83plus/ASCII: Svss04.txt

05. **Licensed Child Care Slots (Single Variable Small Sample *n* < 30)**
The following data represents the number of licensed childcare slots in a random sample of 15 Denver neighborhoods.
Reference: The Piton Foundation, Denver, Colorado

    523 106 184 121 357 319 656 170
    241 226 741 172 266 423 212

    File names   Excel: Svss05.xls
                        Minitab: Svss05.mtp
                        TI-83plus/ASCII: Svss05.txt

06. **Subsidized Housing (Single Variable Small Sample *n* < 30)**
The following data represent the percentage of subsidized housing in a random sample of 14 Denver neighborhoods.
Reference: The Piton Foundation, Denver, Colorado

    10.2 11.8   9.7 22.3   6.8 10.4 11.0
     5.4   6.6 13.7 13.6   6.5 16.0 24.8

    File names   Excel: Svss06.xls
                        Minitab: Svss06.mtp
                    TI-83plus/ASCII: Svss06.txt

07. **Sulfate in Ground Water (Single Variable Small Sample *n* < 30)**
The following data represent naturally occurring amounts of sulfate SO_4 in well water. Units: parts per million. The data is from a random sample of 24 water wells in Northwest Texas.
Reference: Union Carbide Corporation Technical Report K/UR-1

    1850 1150 1340 1325 2500 1060 1220 2325 460
    2000 1500 1775   620 1950   780   840 2650 975
     860   495 1900 1220 2125   990

    File names   Excel: Svss07.xls
                        Minitab: Svss07.mtp
                        TI-83plus/ASCII: Svss07.txt

08. **Earth's Rotation Rate (Single Variable Small Sample *n* < 30)**
The following data represent changes in the earth's rotation (i.e. day length). Units 0.00001 second. The data is for a random sample of 23 years.
Reference: *Acta Astron. Sinica* Vol. 15, p79-85

    -12 110   78 126 -35 104 111   22 -31   92
     51   36 231 -13   65 119   21 104 112 -15
    137 139 101

    File names   Excel: Svss08.xls
                        Minitab: Svss08.mtp
                        TI-83plus/ASCII: Svss08.txt

09. **Blood Glucose (Single Variable Small Sample $n < 30$)**
The following data represent glucose levels (mg/100ml) in the blood for a random sample of 27 non-obese adult subjects.
Reference: *Diabetologia*, Vol. 16, p 17-24

80	85	75	90	70	97	91	85	90	85
105	86	78	92	93	90	80	102	90	90
99	93	91	86	98	86	92			

File names   Excel: Svss09.xls
                             Minitab: Svss09.mtp
                             TI-83plus/ASCII: Svss09.txt

10. **Plant Species (Single Variable Small Sample $n < 30$)**
The following data represent the observed number of native plant species from random samples of study plots on different islands in the Galapagos Island chain.
Reference: *Science*, Vol. 179, p 893-895

23	26	33	73	21	35	30	16	3	17
9	8	9	19	65	12	11	89	81	7
23	95	4	37	28					

File names   Excel: Svss10.xls
                             Minitab: Svss10.mtp
                             TI-83plus/ASCII: Svss10.txt

11. **Apples (Single Variable Small Sample $n < 30$)**
The following data represent mean fruit weight (grams) of apples per tree for a random sample of 28 trees in an agricultural experiment.
Reference: *Aust. J. Agric res.*, Vol. 25, p783-790

85.3	86.9	96.8	108.5	113.8	87.7	94.5	99.9	92.9
67.3	90.6	129.8	48.9	117.5	100.8	94.5	94.4	98.9
96.0	99.4	79.1	108.5	84.6	117.5	70.0	104.4	127.1
135.0								

File names   Excel: Svss11.xls
                             Minitab: Svss11.mtp
                             TI-83plus/ASCII: Svss11.txt

Time Series Data for Control Charts or P Charts
File name prefix: Tscc followed by the number of the data file

01. **Yield of Wheat (Time Series for Control Chart)**
The following data represent annual yield of wheat in tonnes (one ton = 1.016 tonne) for an experimental plot of land at Rothamsted experiment station U.K. over a period of thirty consecutive years.
Reference: Rothamsted Experiment Station U.K.

We will use the following target production values:
target mu = 2.6 tonnes
target sigma = 0.40 tonnes

1.73	1.66	1.36	1.19	2.66	2.14	2.25	2.25	2.36	2.82
2.61	2.51	2.61	2.75	3.49	3.22	2.37	2.52	3.43	3.47
3.20	2.72	3.02	3.03	2.36	2.83	2.76	2.07	1.63	3.02

File names   Excel: Tscc01.xls
                          Minitab: Tscc01.mtp
                          TI-83plus/ASCII: Tscc01.txt

02. **Pepsico Stock Closing Prices (Time Series for Control Chart)**
The following data represent a random sample of 25 weekly closing prices in dollars per share of Pepsico stock for 25 consecutive days.
Reference: *The Denver Post*
The long term estimates for weekly closings are
target mu = 37 dollars per share
target sigma = 1.75 dollars per share

37.000	36.500	36.250	35.250	35.625	36.500	37.000	36.125
35.125	37.250	37.125	36.750	38.000	38.875	38.750	39.500
39.875	41.500	40.750	39.250	39.000	40.500	39.500	40.500
37.875							

File names   Excel: Tscc02.xls
                          Minitab: Tscc02.mtp
                          TI-83plus/ASCII: Tscc02.txt

03.   **Pepsico Stock Volume Of Sales (Time Series for Control Chart)**
The following data represent volume of sales (in hundreds of thousands of shares) of Pepsico stock for 25 consecutive days.
Reference: The Denver Post, business section
For the long term mu and sigma use
  target mu = 15
  target sigma = 4.5

19.00	29.63	21.60	14.87	16.62	12.86	12.25	20.87
23.09	21.71	11.14	5.52	9.48	21.10	15.64	10.79
13.37	11.64	7.69	9.82	8.24	12.11	7.47	12.67
12.33							

File names   Excel: Tscc03.xls
                      Minitab: Tscc03.mtp
                      TI-83plus/ASCII: Tscc03.txt

04.   **Futures Quotes For The Price Of Coffee Beans (Time Series for Control Chart)**
The following data represent futures options quotes for the price of coffee beans (dollars per pound) for 20 consecutive business days.
Use the following estimated target values for pricing
  target mu = $2.15
  target sigma = $0.12

2.300	2.360	2.270	2.180	2.150	2.180	2.120	2.090	2.150	2.200
2.170	2.160	2.100	2.040	1.950	1.860	1.910	1.880	1.940	1.990

File names   Excel: Tscc04.xls
                      Minitab: Tscc04.mtp
                      TI-83plus/ASCII: Tscc04.txt

05.   **Incidence Of Melanoma Tumors (Time Series for Control Chart)**
The following data represent number of cases of melanoma skin cancer (per 100,000 population) in Connecticut for each of the years 1953 to 1972.
Reference: *Inst. J. Cancer* , Vol. 25, p95-104
Use the following long term values (mu and sigma)
  target mu = 3
  target sigma = 0.9

2.4	2.2	2.9	2.5	2.6	3.2	3.8	4.2	3.9	3.7
3.3	3.7	3.9	4.1	3.8	4.7	4.4	4.8	4.8	4.8

File names   Excel: Tscc05.xls
                      Minitab: Tscc05.mtp
                      TI-83plus/ASCII: Tscc05.txt

06. Percent Change In Consumer Price Index (Time Series for Control Chart)
The following data represent annual percent change in consumer price index for a sequence of recent years.
Reference: *Statistical Abstract Of The United States*
Suppose an economist recommends the following long term target values for mu and sigma.
  target mu = 4.0%
  target sigma = 1.0%

```
1.3   1.3   1.6   2.9   3.1   4.2    5.5   5.7   4.4   3.2
6.2  11.0   9.1   5.8   6.5   7.6   11.3  13.5  10.3   6.2
3.2   4.3   3.6   1.9   3.6   4.1    4.8   5.4   4.2   3.0
```

File names   Excel: Tscc06.xls
                        Minitab: Tscc06.mtp
                        TI-83plus/ASCII: Tscc06.txt

07. Broken Eggs (Time Series for *P* Chart)
The following data represent the number of broken eggs in a case of 10 dozen eggs (120 eggs). The data represent 21 days or 3 weeks of deliveries to a small grocery store.

```
14   23   18    9   17   14   12   11   10   17
12   25   18   15   19   22   14   22   15   10
13
```

File names   Excel: Tscc07.xls
                        Minitab: Tscc07.mtp
                        TI-83plus/ASCII: Tscc07.txt

08. Theater Seats (Time Series for *P* Chart)
The following data represent the number of empty seats at each show of a Community Theater production. The theater has 325 seats. The show ran 18 times.

```
28   19   41   38   32   47   53   17   29
32   31   27   25   33   26   62   15   12
```

File names   Excel: Tscc08.xls
                        Minitab: Tscc08.mtp
                        TI-83plus/ASCII: Tscc08.txt

09. Rain (Time Series for *P* Chart)
The following data represents the number of rainy days at Waikiki beach, Hawaii during the prime tourist season of December and January (62 days). The data was taken over a 20-year period.

```
21   27   19   17    6    9   25   36   23   26
12   16   27   41   18    8   10   22   15   24
```

File names   Excel: Tscc09.xls
                        Minitab: Tscc09.mtp
                        TI-83plus/ASCII: Tscc09.txt

10. **Quality Control (Time Series for *P* Chart)**
    The following data represent the number of defective toys in a case of 500 toys coming off a production line. Every day for 35 consecutive days, a case was selected at random.

26	23	33	49	28	42	29	41	27	25
35	21	48	12	5	15	36	55	13	16
93	8	38	11	39	18	7	33	29	42
26	19	47	53	61					

    File names   Excel: Tscc10.xls
                                   Minitab: Tscc10.mtp
                                   TI-83plus/ASCII: Tscc10.txt

Two Variable Independent Samples
File name prefix: Tvis followed by the number of the data file

01.   Heights of Football Players Versus Heights of Basketball Players
(Two variable independent large samples)
The following data represent heights in feet of 45 randomly selected pro football players, and 40 randomly selected pro basketball players.
Reference: Sports Encyclopedia Pro Football, and Official NBA Basketball Encyclopedia

$X1$ = heights (ft.) of pro football players

6.33	6.50	6.50	6.25	6.50	6.33	6.25	6.17	6.42	6.33
6.42	6.58	6.08	6.58	6.50	6.42	6.25	6.67	5.91	6.00
5.83	6.00	5.83	5.08	6.75	5.83	6.17	5.75	6.00	5.75
6.50	5.83	5.91	5.67	6.00	6.08	6.17	6.58	6.50	6.25
6.33	5.25	6.67	6.50	5.83					

$X2$ = heights (ft.) of pro basketball players

6.08	6.58	6.25	6.58	6.25	5.92	7.00	6.41	6.75	6.25
6.00	6.92	6.83	6.58	6.41	6.67	6.67	5.75	6.25	6.25
6.50	6.00	6.92	6.25	6.42	6.58	6.58	6.08	6.75	6.50
6.83	6.08	6.92	6.00	6.33	6.50	6.58	6.83	6.50	6.58

File names   Excel: Tvis01.xls
                      Minitab: Tvis02.mtp
                      TI-83plus/ASCII: X1 data is stored in Tvis01L1.txt
                                        X2 data is stored in Tvis01L2.txt

02.   Petal Length for Iris Virginica Versus Petal Length for Iris Setosa
(Two variable independent large samples)
The following data represent petal length (cm.) for a random sample of 35 iris virginica and a random sample of 38 iris setosa
Reference: Anderson, E., *Bull. Amer. Iris Soc.*

$X1$ = petal length (c.m.) iris virginica

5.1	5.8	6.3	6.1	5.1	5.5	5.3	5.5	6.9	5.0	4.9	6.0	4.8	6.1	5.6	5.1
5.6	4.8	5.4	5.1	5.1	5.9	5.2	5.7	5.4	4.5	6.1	5.3	5.5	6.7	5.7	4.9
4.8	5.8	5.1													

$X2$ = petal length (c.m.) iris setosa

1.5	1.7	1.4	1.5	1.5	1.6	1.4	1.1	1.2	1.4	1.7	1.0	1.7	1.9	1.6	1.4
1.5	1.4	1.2	1.3	1.5	1.3	1.6	1.9	1.4	1.6	1.5	1.4	1.6	1.2	1.9	1.5
1.6	1.4	1.3	1.7	1.5	1.7										

File names   Excel: Tvis02.xls
                      Minitab: Tvis02.mtp
                      TI-83plus/ASCII: X1 data is stored in Tvis02L1.txt
                                        X2 data is stored in Tvis02L2.txt

03. **Sepal Width Of Iris Versicolor Versus Iris Virginica**
(Two variable independent larage samples)

The following data represent sepal width (cm.) for a random sample of 40 iris versicolor and a random sample of 42 iris virginica

Reference: Anderson, E., *Bull. Amer. Iris Soc.*

X1 = sepal width (c.m.) iris versicolor
3.2 3.2 3.1 2.3 2.8 2.8 3.3 2.4 2.9 2.7 2.0 3.0 2.2 2.9 2.9 3.1
3.0 2.7 2.2 2.5 3.2 2.8 2.5 2.8 2.9 3.0 2.8 3.0 2.9 2.6 2.4 2.4
2.7 2.7 3.0 3.4 3.1 2.3 3.0 2.5

X2 = sepal width (c.m.) iris virginica
3.3 2.7 3.0 2.9 3.0 3.0 2.5 2.9 2.5 3.6 3.2 2.7 3.0 2.5 2.8 3.2
3.0 3.8 2.6 2.2 3.2 2.8 2.8 2.7 3.3 3.2 2.8 3.0 2.8 3.0 2.8 3.8
2.8 2.8 2.6 3.0 3.4 3.1 3.0 3.1 3.1 3.1

File names	Excel: Tvis03.xls
	Minitab: Tvis03.mtp
	TI-83plus/ASCII:X1 data is stored in Tvis03L1.txt
	X2 data is stored in Tvis03L2.txt

04. **Archaeology, Ceramics (Two variable independent large samples)**

The following data represent independent random samples of shard counts of painted ceramics found at the Wind Mountain archaeological site.

Reference: Woosley and McIntyre, *Mimbres Mogollon Archaeology*, Univ. New Mexico Press

X1 = count Mogollon red on brown
52	10	8	71	7	31	24	20	17	5
16	75	25	17	14	33	13	17	12	19
67	13	35	14	3	7	9	19	16	22
7	10	9	49	6	13	24	45	14	20
3	6	30	41	26	32	14	33	1	48
44	14	16	15	13	8	61	11	12	16
20	39								

X2 = count Mimbres black on white
61	21	78	9	14	12	34	54	10	15
43	9	7	67	18	18	24	54	8	10
16	6	17	14	25	22	25	13	23	12
36	10	56	35	79	69	41	36	18	25
27	27	11	13						

File names	Excel: Tvis04.xls
	Minitab: Tvis04.mtp
	TI-83plus/ASCII:X1 data is stored in Tvis04L1.txt
	X2 data is stored in Tvis04L2.txt

05. Agriculture, Water Content of Soil (Two variable independent large samples)

The following data represent soil water content (% water by volume) for independent random samples of soil from two experimental fields growing bell peppers.

Reference: *Journal of Agricultural, Biological, and Environmental Statistics*, Vol. 2, No. 2,
        p 149-155

X_1 = soil water content from field I

15.1	11.2	10.3	10.8	16.6	8.3	9.1	12.3	9.1	14.3
10.7	16.1	10.2	15.2	8.9	9.5	9.6	11.3	14.0	11.3
15.6	11.2	13.8	9.0	8.4	8.2	12.0	13.9	11.6	16.0
9.6	11.4	8.4	8.0	14.1	10.9	13.2	13.8	14.6	10.2
11.5	13.1	14.7	12.5	10.2	11.8	11.0	12.7	10.3	10.8
11.0	12.6	10.8	9.6	11.5	10.6	11.7	10.1	9.7	9.7
11.2	9.8	10.3	11.9	9.7	11.3	10.4	12.0	11.0	10.7
8.8	11.1								

X_2 = soil water content from field II

12.1	10.2	13.6	8.1	13.5	7.8	11.8	7.7	8.1	9.2
14.1	8.9	13.9	7.5	12.6	7.3	14.9	12.2	7.6	8.9
13.9	8.4	13.4	7.1	12.4	7.6	9.9	26.0	7.3	7.4
14.3	8.4	13.2	7.3	11.3	7.5	9.7	12.3	6.9	7.6
13.8	7.5	13.3	8.0	11.3	6.8	7.4	11.7	11.8	7.7
12.6	7.7	13.2	13.9	10.4	12.8	7.6	10.7	10.7	10.9
12.5	11.3	10.7	13.2	8.9	12.9	7.7	9.7	9.7	11.4
11.9	13.4	9.2	13.4	8.8	11.9	7.1	8.5	14.0	14.2

File names   Excel: Tvis05.xls
                  Minitab: Tvis05.mtp
                  TI-83plus/ASCII: X_1 data is stored in Tvis05L1.txt
                              X_2 data is stored in Tvis05L2.txt

06. Rabies (Two variable independent small samples)

The following data represent number of cases of red fox rabies for a random sample of 16 areas in each of two different regions of southern Germany.

Reference: Sayers, B., *Medical Informatics*, Vol. 2, 11-34

X_1 = number cases in region 1

10 2 2 5 3 4 3 3 4 0 2 6 4 8 7 4

X_2 = number cases in region 2

1 1 2 1 3 9 2 2 4 5 4 2 2 0 0 2

File names   Excel: Tvis06.xls
                  Minitab: Tvis06.mtp
                  TI-83plus/ASCII: X_1 data is stored in Tvis06L1.txt
                              X_2 data is stored in Tvis06L2.txt

**07. Weight of Football Players Versus Weight of Basketball Players
(Two variable independent small samples)**
The following data represent weights in pounds of 21 randomly selected pro football players, and 19 randomly selected pro basketball players.
Reference: *Sports Encyclopedia of Pro Football*, and *Official NBA Basketball Encyclopedia*

X1 = weights (lb) of pro football players

245	262	255	251	244	276	240	265	257	252	282
256	250	264	270	275	245	275	253	265	270	

X2 = weights (lb) of pro basketball

205	200	220	210	191	215	221	216	228	207
225	208	195	191	207	196	181	193	201	

File names

Excel: Tvis07.xls
Minitab: Tvis07.mtp
TI-83plus/ASCII: X1 data is stored in Tvis07L1.txt
X2 data is stored in Tvis07L2.txt

08. Birth Rate (Two variable independent small samples)
The following data represent birth rate (per 1000 residential population) for independent random samples of counties in California and Maine.
Reference: *County and City Data Book* 12th edition, U.S. Dept. of Commerce

X1 = birth rate in California counties

14.1	18.7	20.4	20.7	16.0	12.5	12.9	9.6	17.6
18.1	14.1	16.6	15.1	18.5	23.6	19.9	19.6	14.9
17.7	17.8	19.1	22.1	15.6				

X2 = birth rate in Maine counties

15.1	14.0	13.3	13.8	13.5	14.2	14.7	11.8	13.5	13.8
16.5	13.8	13.2	12.5	14.8	14.1	13.6	13.9	15.8	

File names

Excel: Tvis08.xls
Minitab: Tvis08.mtp
TI-83plus/ASCII: X1 data is stored in Tvis08L1.txt
X2 data is stored in Tvis08L2.txt

09. Death Rate (Two variable independent small samples)
The following data represents death rate (per 1000 resident population) for independent random samples of counties in Alaska and Texas.
Reference: *County and City Data Book* 12th edition, U.S. Dept. of Commerce

X1 = death rate in Alaska counties
  1.4   4.2   7.3   4.8   3.2   3.4   5.1   5.4
  6.7   3.3   1.9   8.3   3.1   6.0   4.5   2.5

X2 = death rate in Texas counties
  7.2   5.8   10.5   6.6   6.9   9.5   8.6   5.9   9.1
  5.4   8.8   6.1   9.5   9.6   7.8   10.2   5.6   8.6

File names	Excel: Tvis09.xls
	Minitab: Tvis09.mtp
	TI-83plus/ASCII:X1 data is stored in Tvis09L1.txt
	X2 data is stored in Tvis09L2.txt

10. Pickup Trucks (Two variable independent small samples)
The following data represents retail price (in thousands of dollars) for independent random samples of models of pickup trucks.
Reference: *Consumer Guide* Vol.681

X1 = prices for different GMC Sierra 1500 models
  17.4   23.3   29.2   19.2   17.6   19.2   23.6   19.5   22.2
  24.0   26.4   23.7   29.4   23.7   26.7   24.0   24.9

X2 = prices for different Chevrolet Silverado 1500 models
  17.5   23.7   20.8   22.5   24.3   26.7   24.5   17.8
  29.4   29.7   20.1   21.1   22.1   24.2   27.4   28.1

File names	Excel: Tvis10.xls
	Minitab: Tvis10.mtp
	TI-83plus/ASCII:X1 data is stored in Tvis10L1.txt
	X2 data is stored in Tvis10L2.txt

Two Variable Dependent Samples
File name prefix: Tvds followed by the number of the data file

01. **Average Faculty Salary, Males vs Female (Two variable dependent samples)**
In following data pairs
  A = average salaries for males ($1000/yr)
  B = average salaries for females ($1000/yr)
for assistant professors at the same college or university. A random sample of 22 US. colleges and universities was used.
Reference: *Academe, Bulletin of the American Association of University Professors*

A: 34.5 30.5 35.1 35.7 31.5 34.4 32.1 30.7 33.7 35.3
B: 33.9 31.2 35.0 34.2 32.4 34.1 32.7 29.9 31.2 35.5

A: 30.7 34.2 39.6 30.5 33.8 31.7 32.8 38.5 40.5 25.3
B: 30.2 34.8 38.7 30.0 33.8 32.4 31.7 38.9 41.2 25.5

A: 28.6 35.8
B: 28.0 35.1

File names   Excel: Tvds01.xls
               Minitab: Tvds01.mtp
               TI-83plus/ASCII: A data is stored in Tvds01L1.txt
                                B data is stored in Tvds01L2.txt

02. **Unemployment for College Graduates Versus High School Only**
(Two variable dependent samples)
In the following data pairs
  A = Percent unemployment for college graduates
  B = Percent unemployment for high school only graduates
The data are paired by year
Reference: *Statistical Abstract of the United States*

A: 2.8 2.2 2.2 1.7 2.3 2.3 2.4 2.7 3.5 3.0 1.9 2.5
B: 5.9 4.9 4.8 5.4 6.3 6.9 6.9 7.2 10.0 8.5 5.1 6.9

File names   Excel: Tvds02.xls
               Minitab: Tvds02.mtp
               TI-83plus/ASCII: A data is stored in Tvds02L1.txt
                                B data is stored in Tvds02L2.txt

03. Number of Navajo Hogans versus Modern Houses (Two variable dependent samples)
In the following data pairs
     A = Number of traditional Navajo hogans in a given district
     B = Number of modern houses in a given district
The data are paired by districts on the Navajo reservation. A random sample of 8 districts were used.
Reference: *Navajo Architecture, Forms, History, Distributions* by S.C. Jett and V.E. Spencer,
         Univ. of Arizona Press

A:	13	14	46	32	15	47	17	18
B:	18	16	68	9	11	28	50	50

File names	Excel: Tvds03.xls
	Minitab: Tvds03.mtp
	TI-83plus/ASCII: A data is stored in Tvds03L1.txt
	B data is stored in Tvds03L2.txt

04. Temperatures in Miami versus Honolulu (Two variable dependent samples)

In the following data pairs
     A = Average monthly temperature in Miami
     B = Average monthly temperature in Honolulu
The data are paired by month.
Reference: U.S. Department of Commerce Environmental Data Service

A:	67.5	68.0	71.3	74.9	78.0	80.9	82.2	82.7	81.6	77.8	72.3	68.5
B:	74.4	72.6	73.3	74.7	76.2	78.0	79.1	79.8	79.5	78.4	76.1	73.7

File names	Excel: Tvds04.xls
	Minitab: Tvds04.mtp
	TI-83plus/ASCII: A data is stored in Tvds04L1.txt
	B data is stored in Tvds04L2.txt

05. January/February Ozone Column (Two variable dependent samples)
In the following pairs the data represents the thickness of the ozone column in Dobson units: one millimeter ozone at standard temperature and pressure.
     A = monthly mean thickness in January
     B = monthly mean thickness in February
The data are paired by year for a random sample of 15 years.
Reference: Laboratorium für Atmospharensphysic, Switzerland

A:	360	324	377	336	383	361	369	349
B:	365	325	359	352	397	351	367	397

A:	301	354	344	329	337	387	378
B:	335	338	349	393	370	400	411

File names	Excel: Tvds05.xls
	Minitab: Tvds05.mtp
	TI-83plus/ASCII: A data is stored in Tvds05L1.txt
	B data is stored in Tvds05L2.txt

06. Birth Rate/Death Rate **(Two variable dependent samples)**
In the following data pairs,
A = birth rate (per 1000 resident population)
B = death rate (per 1000 resident population)
The data are paired by county in Iowa
Reference: *County and City Data Book* 12th edition, U.S. Dept. of Commerce

A: 12.7 13.4 12.8 12.1 11.6 11.1 14.2
B: 9.8 14.5 10.7 14.2 13.0 12.9 10.9

A: 12.5 12.3 13.1 15.8 10.3 12.7 11.1
B: 14.1 13.6   9.1 10.2 17.9 11.8   7.0

File names	Excel: Tvds06.xls
	Minitab: Tvds06.mtp
	TI-83plus/ASCII:A data is stored in Tvds06L1.txt
	B data is stored in Tvds06L2.txt

07. Democrat/Republican (Two variable dependent samples)
In the following data pairs
A = percentage of voters who voted Democrat
B = percentage of voters who voted Republican
in a recent national election. The data are paired by county in Indiana
Reference: *County and City Data Book* 12th edition, U.S. Dept. of Commerce

A: 42.2 34.5 44.0 34.1 41.8 40.7 36.4 43.3 39.5
B: 35.4 45.8 39.4 40.0 39.2 40.2 44.7 37.3 40.8

A: 35.4 44.1 41.0 42.8 40.8 36.4 40.6 37.4
B: 39.3 36.8 35.5 33.2 38.3 47.7 41.1 38.5

File names	Excel: Tvds07.xls
	Minitab: Tvds07.mtp
	TI-83plus/ASCII:A data is stored in Tvds07L1.txt
	B data is stored in Tvds07L2.txt

08. Santiago Pueblo Pottery (Two variable dependent samples)
In the following data
A = percentage of utility pottery
B = percentage of ceremonial pottery
found at the Santiago Pueblo archaeological site. The data are paired by location of discovery.
Reference: Laboratory of Anthropology, Notes 475, Santa Fe, New Mexico

A: 41.4 49.6 55.6 49.5 43.0 54.6 46.8 51.1 43.2 41.4
B: 58.6 50.4 44.4 59.5 57.0 45.4 53.2 48.9 56.8 58.6

File names	Excel: Tvds08.xls
	Minitab: Tvds08.mtp
	TI-83plus/ASCII:A data is stored in Tvds08L1.txt
	B data is stored in Tvds08L2.txt

09. Poverty Level (Two variable dependent samples)

In the following data pairs

    A = percentage of population below poverty level in 1998

    B = percentage of population below poverty level in 1990

The data are grouped by state and District of Columbia

Reference: *Statistical Abstract of the United States*, 120th edition

A:	14.5	9.4	16.6	14.8	15.4	9.2	9.5	10.3	22.3	13.1	
B:	19.2	11.4	13.7	19.6	13.9	13.7	6.0	6.9	21.1	14.4	

A:	13.6	10.9	13.0	10.1	9.4	9.1	9.6	13.5	19.1	10.4	
B:	15.8	11.0	14.9	13.7	13.0	10.4	10.3	17.3	23.6	13.1	

A:	7.2	8.7	11.0	10.4	17.6	9.8	16.6	12.3	10.6	9.8	
B:	9.9	10.7	14.3	12.0	25.7	13.4	16.3	10.3	9.8	6.3	

A:	8.6	20.4	16.7	14.0	15.1	11.2	14.1	15.0	11.2	11.6	
B:	9.2	20.9	14.3	13.0	13.7	11.5	15.6	9.2	11.0	7.5	

A:	13.7	10.8	13.4	15.1	9.0	9.9	8.8	8.9	17.8	8.8	10.6
B:	16.2	13.3	16.9	15.9	8.2	10.9	11.1	8.9	18.1	9.3	11.0

File names   Excel: Tvds09.xls

                       Minitab: Tvds09.mtp

                       TI-83plus/ASCII: A data is stored in Tvds09L1.txt

                                       B data is stored in Tvds09L2.txt

10. Cost of Living Index (Two variable dependent samples)

The following data pairs represent cost of living index for

    A = grocery items

    B = health care

The data are grouped by metropolitan areas.

Reference: *Statistical Abstract of the United States*, 120th edition

Grocery

A:	96.6	97.5	113.9	88.9	108.3	99.0	97.3	87.5	96.8
B:	91.6	95.9	114.5	93.6	112.7	93.6	99.2	93.2	105.9

A:	102.1	114.5	100.9	100.0	100.7	99.4	117.1	111.3	102.2
B:	110.8	127.0	91.5	100.5	104.9	104.8	124.1	124.6	109.1

A:	95.3	91.1	95.7	87.5	91.8	97.9	97.4	102.1	94.0
B:	98.7	95.8	99.7	93.2	100.7	96.0	99.6	98.4	94.0

A:	115.7	118.3	101.9	88.9	100.7	99.8	101.3	104.8	100.9
B:	121.2	122.4	110.8	81.2	104.8	109.9	103.5	113.6	94.6

A:	102.7	98.1	105.3	97.2	105.2	108.1	110.5	99.3	99.7
B:	109.8	97.6	109.8	107.4	97.7	124.2	110.9	106.8	94.8

File names   Excel: Tvds10.xls

                        Minitab: Tvds10.mtp

                        TI-83plus/ASCII: A data is stored in Tvds10L1.txt

                                          B data is stored in Tvds10L2.txt

Simple Linear Regression
File name prefix: Slr followed by the number of the data file

01. List Price versus Best Price for a New GMC Pickup Truck (Simple Linear Regression)
In the following data
$\quad$ X = List price (in $1000) for a GMC pickup truck
$\quad$ Y = Best price (in $1000) for a GMC pickup truck
Reference: *Consumer's Digest*

X:	12.4	14.3	14.5	14.9	16.1	16.9	16.5	15.4	17.0	17.9
Y:	11.2	12.5	12.7	13.1	14.1	14.8	14.4	13.4	14.9	15.6

X:	18.8	20.3	22.4	19.4	15.5	16.7	17.3	18.4	19.2	17.4
Y:	16.4	17.7	19.6	16.9	14.0	14.6	15.1	16.1	16.8	15.2

X:	19.5	19.7	21.2
Y:	17.0	17.2	18.6

File names$\qquad$Excel: Slr01.xls
$\qquad\qquad\qquad\qquad$Minitab: Slr01.mtp
$\qquad\qquad\qquad\qquad$TI-83plus/ASCII: X data is stored in Slr01L1.txt
$\qquad\qquad\qquad\qquad\qquad\qquad\quad$Y data is stored in Slr01L2.txt

02. Cricket Chirps versus Temperature (Simple Linear Regression)
In the following data
$\quad$ X = chirps/sec for the striped ground cricket
$\quad$ Y = temperature in degrees Fahrenheit
Reference: *The Song of Insects* by Dr.G.W. Pierce, Harvard College Press

X:	20.0	16.0	19.8	18.4	17.1	15.5	14.7	17.1
Y:	88.6	71.6	93.3	84.3	80.6	75.2	69.7	82.0

X:	15.4	16.2	15.0	17.2	16.0	17.0	14.4
Y:	69.4	83.3	79.6	82.6	80.6	83.5	76.3

File names$\qquad$Excel: Slr02.xls
$\qquad\qquad\qquad\qquad$Minitab: Slr02.mtp
$\qquad\qquad\qquad\qquad$TI-83plus/ASCII: X data is stored in Slr02L1.txt
$\qquad\qquad\qquad\qquad\qquad\qquad\quad$Y data is stored in Slr02L2.txt

03. **Diameter of Sand Granules versus Slope on Beach (Simple Linear Regression)**
In the following data pairs
X = median diameter (mm) of granules of sand
Y = gradient of beach slope in degrees
The data is for naturally occurring ocean beaches
Reference: *Physical geography* by A.M King, Oxford Press, England

X:	0.170	0.190	0.220	0.235	0.235	0.300	0.350	0.420	0.850
Y:	0.630	0.700	0.820	0.880	1.150	1.500	4.400	7.300	11.300

File names   Excel: Slr03.xls
Minitab: Slr03.mtp
TI-83plus/ASCII: X data is stored in Slr03L1.txt
Y data is stored in Slr03L2.txt

04. **National Unemployment Male versus Female (Simple Linear Regression)**
In the following data pairs
X = national unemployment rate for adult males
Y = national unemployment rate for adult females
Reference: *Statistical Abstract of the United States*

X:	2.9	6.7	4.9	7.9	9.8	6.9	6.1	6.2	6.0	5.1	4.7	4.4	5.8
Y:	4.0	7.4	5.0	7.2	7.9	6.1	6.0	5.8	5.2	4.2	4.0	4.4	5.2

File names   Excel: Slr04.xls
Minitab: Slr04.mtp
TI-83plus/ASCII: X data is stored in Slr04L1.txt
Y data is stored in Slr04L2.txt

05.   Fire and Theft in Chicago (Simple Linear Regression)
In the following data pairs
    X = fires per 1000 housing units
     Y = thefts per 1000 population
within the same Zip code in the Chicago metro area
Reference: U.S. Commission on Civil Rights

X:	6.2	9.5	10.5	7.7	8.6	34.1	11.0	6.9	7.3	15.1
Y:	29	44	36	37	53	68	75	18	31	25

X:	29.1	2.2	5.7	2.0	2.5	4.0	5.4	2.2	7.2	15.1
Y:	34	14	11	11	22	16	27	9	29	30

X:	16.5	18.4	36.2	39.7	18.5	23.3	12.2	5.6	21.8	21.6
Y:	40	32	41	147	22	29	46	23	4	31

X:	9.0	3.6	5.0	28.6	17.4	11.3	3.4	11.9	10.5	10.7
Y:	39	15	32	27	32	34	17	46	42	43

X:	10.8	4.8
Y:	34	19

File names   Excel: Slr05.xls
                              Minitab: Slr05.mtp
                              TI-83plus/ASCII: X data is stored in Slr05L1.txt
                                                      Y data is stored in Slr05L2.txt

06. Auto Insurance in Sweden (Simple Linear Regression)

In the following data

X = number of claims

Y = total payment for all the claims in thousands of Swedish Kronor

for geographical zones in Sweden

Reference: Swedish Committee on Analysis of Risk Premium in Motor Insurance

X:	108	19	13	124	40	57	23	14	45	10
Y:	392.5	46.2	15.7	422.2	119.4	170.9	56.9	77.5	214.0	65.3

X:	5	48	11	23	7	2	24	6	3	23
Y:	20.9	248.1	23.5	39.6	48.8	6.6	134.9	50.9	4.4	113.0

X:	6	9	9	3	29	7	4	20	7	4
Y:	14.8	48.7	52.1	13.2	103.9	77.5	11.8	98.1	27.9	38.1

X:	0	25	6	5	22	11	61	12	4	16
Y:	0.0	69.2	14.6	40.3	161.5	57.2	217.6	58.1	12.6	59.6

X:	13	60	41	37	55	41	11	27	8	3
Y:	89.9	202.4	181.3	152.8	162.8	73.4	21.3	92.6	76.1	39.9

X:	17	13	13	15	8	29	30	24	9	31
Y:	142.1	93.0	31.9	32.1	55.6	133.3	194.5	137.9	87.4	209.8

X:	14	53	26
Y:	95.5	244.6	187.5

File names   Excel: Slr06.xls

Minitab: Slr06.mtp

TI-83plus/ASCII: X data is stored in Slr06L1.txt

Y data is stored in Slr06L2.txt

07. Gray Kangaroos (Simple Linear Regression)

In the following data pairs

X = nasal length (mm ×10)

Y = nasal width (mm × 10)

for a male gray kangaroo from a random sample of such animals.

Reference: *Australian Journal of Zoology*, Vol. 28, p607-613

X:	609	629	620	564	645	493	606	660	630	672
Y:	241	222	233	207	247	189	226	240	215	231

X:	778	616	727	810	778	823	755	710	701	803
Y:	263	220	271	284	279	272	268	278	238	255

X:	855	838	830	864	635	565	562	580	596	597
Y:	308	281	288	306	236	204	216	225	220	219

X:	636	559	615	740	677	675	629	692	710	730
Y:	201	213	228	234	237	217	211	238	221	281

X:	763	686	717	737	816
Y:	292	251	231	275	275

File names   Excel: Slr07.xls

                        Minitab: Slr07.mtp

                        TI-83plus/ASCII: X data is stored in Slr07L1.txt

                                         Y data is stored in Slr07L2.txt

08. Pressure and Weight in Cryogenic Flow Meters (Simple Linear Regression)

In the following data pairs

X = pressure (lb/sq in) of liquid nitrogen

Y = weight in pounds of liquid nitrogen passing through flow meter each second

Reference: *Technometrics*, Vol. 19, p353-379

X:	75.1	74.3	88.7	114.6	98.5	112.0	114.8	62.2	107.0
Y:	577.8	577.0	570.9	578.6	572.4	411.2	531.7	563.9	406.7

X:	90.5	73.8	115.8	99.4	93.0	73.9	65.7	66.2	77.9
Y:	507.1	496.4	505.2	506.4	510.2	503.9	506.2	506.3	510.2

X:	109.8	105.4	88.6	89.6	73.8	101.3	120.0	75.9	76.2
Y:	508.6	510.9	505.4	512.8	502.8	493.0	510.8	512.8	513.4

X:	81.9	84.3	98.0
Y:	510.0	504.3	522.0

File names   Excel: Slr08.xls

                        Minitab: Slr08.mtp

                        TI-83plus/ASCII: X data is stored in Slr08L1.txt

                                         Y data is stored in Slr07L2.txt

09. Ground Water Survey (Simple Linear Regression)

In the following data

X = pH of well water

Y = Bicarbonate (parts per million) of well water

The data is by water well from a random sample of wells in Northwest Texas.

Reference: Union Carbide Technical Report K/UR-1

X:	7.6	7.1	8.2	7.5	7.4	7.8	7.3	8.0	7.1	7.5
Y:	157	174	175	188	171	143	217	190	142	190

X:	8.1	7.0	7.3	7.8	7.3	8.0	8.5	7.1	8.2	7.9
Y:	215	199	262	105	121	81	82	210	202	155

X:	7.6	8.8	7.2	7.9	8.1	7.7	8.4	7.4	7.3	8.5
Y:	157	147	133	53	56	113	35	125	76	48

X:	7.8	6.7	7.1	7.3
Y:	147	117	182	87

File names   Excel: Slr09.xls

Minitab: Slr09.mtp

TI-83plus/ASCII: X data is stored in Slr09L1.txt

Y data is stored in Slr09L2.txt

10. Iris Setosa (Simple Linear Regression)

In the following data

X = sepal width (cm)

Y = sepal length (cm)

The data is for a random sample of the wild flower iris setosa.

Reference: Fisher, R.A., *Ann. Eugenics,* Vol. 7 Part II, p 179-188

X:	3.5	3.0	3.2	3.1	3.6	3.9	3.4	3.4	2.9	3.1
Y:	5.1	4.9	4.7	4.6	5.0	5.4	4.6	5.0	4.4	4.9

X:	3.7	3.4	3.0	4.0	4.4	3.9	3.5	3.8	3.8	3.4
Y:	5.4	4.8	4.3	5.8	5.7	5.4	5.1	5.7	5.1	5.4

X:	3.7	3.6	3.3	3.4	3.0	3.4	3.5	3.4	3.2	3.1
Y:	5.1	4.6	5.1	4.8	5.0	5.0	5.2	5.2	4.7	4.8

X:	3.4	4.1	4.2	3.1	3.2	3.5	3.6	3.0	3.4	3.5
Y:	5.4	5.2	5.5	4.9	5.0	5.5	4.9	4.4	5.1	5.0

X:	2.3	3.2	3.5	3.8	3.0	3.8	3.7	3.3
Y:	4.5	4.4	5.0	5.1	4.8	4.6	5.3	5.0

File names   Excel: Slr10.xls

Minitab: Slr10.mtp

TI-83plus/ASCII: X data is stored in Slr10L1.txt

Y data is stored in Slr10L2.txt

11. Pizza Franchise (Simple Linear Regression)
In the following data
   X = annual franchise fee ($1000)
   Y = start up cost ($1000)
for a pizza franchise
Reference: *Business Opportunity Handbook*

X:	25.0	8.5	35.0	15.0	10.0	30.0	10.0	50.0	17.5	16.0
Y:	125	80	330	58	110	338	30	175	120	135

X:	18.5	7.0	8.0	15.0	5.0	15.0	12.0	15.0	28.0	20.0
Y:	97	50	55	40	35	45	75	33	55	90

X:	20.0	15.0	20.0	25.0	20.0	3.5	35.0	25.0	8.5	10.0
Y:	85	125	150	120	95	30	400	148	135	45

X:	10.0	25.0
Y:	87	150

File names	Excel: Slr11.xls
	Minitab: Slr11.mtp
	TI-83plus/ASCII: X data is stored in Slr11L1.txt
	Y data is stored in Slr11L2.txt

12. Prehistoric Pueblos (Simple Linear Regression)
In the following data
   X = estimated year of initial occupation
   Y = estimated year of end of occupation
The data are for each prehistoric pueblo in a random sample of such pueblos in Utah, Arizona, and Nevada.
Reference *Prehistoric Pueblo World* by A. Adler, Univ. of Arizona Press

X:	1000	1125	1087	1070	1100	1150	1250	1150	1100
Y:	1050	1150	1213	1275	1300	1300	1400	1400	1250

X:	1350	1275	1375	1175	1200	1175	1300	1260	1330
Y:	1830	1350	1450	1300	1300	1275	1375	1285	1400

X:	1325	1200	1225	1090	1075	1080	1080	1180	1225
Y:	1400	1285	1275	1135	1250	1275	1150	1250	1275

X:	1175	1250	1250	750	1125	700	900	900	850
Y:	1225	1280	1300	1250	1175	1300	1250	1300	1200

File names	Excel: Slr12.xls
	Minitab: Slr12.mtp
	TI-83plus/ASCII: X data is stored in Slr12L1.txt
	Y data is stored in Slr12L2.txt

Multiple Linear Regression
File name prefix: Mlr followed by the number of the data file

01. Thunder Basin Antelope Study (Multiple Linear Regression)

The data (X1, X2, X3, X4) are for each year.

X1 = spring fawn count/100

X2 = size of adult antelope population/100

X3 = annual precipitation (inches)

X4 = winter severity index (1=mild , 5=severe)

X1	X2	X3	X4
2.90	9.20	13.20	2.00
2.40	8.70	11.50	3.00
2.00	7.20	10.80	4.00
2.30	8.50	12.30	2.00
3.20	9.60	12.60	3.00
1.90	6.80	10.60	5.00
3.40	9.70	14.10	1.00
2.10	7.90	11.20	3.00

File names   Excel: Mlr01.xls

Minitab: Mlr01.mtp

TI-83plus/ASCII:X1 data is stored in Mlr01L1.txt

X2 data is stored in Mlr01L2.txt

X3 data is stored in Mlr01L3.txt

X4 data is stored in Mlr01L4.txt

02. Section 10.5, problem #3 Systolic Blood Pressure Data (Multiple Linear Regression)

The data (X1, X2, X3) are for each patient.

X1 = systolic blood pressure

X2 = age in years

X3 = weight in pounds

X1	X2	X3
132.00	52.00	173.00
143.00	59.00	184.00
153.00	67.00	194.00
162.00	73.00	211.00
154.00	64.00	196.00
168.00	74.00	220.00
137.00	54.00	188.00
149.00	61.00	188.00
159.00	65.00	207.00
128.00	46.00	167.00
166.00	72.00	217.00

File names   Excel: Mlr02.xls

Minitab: Mlr02.mtp

TI-83plus/ASCII:X1 data is stored in Mlr02L1.txt

X2 data is stored in Mlr02L2.txt

X3 data is stored in Mlr02L3.txt

03. Section 10.5, Problem #4 Test Scores for General Psychology (Multiple Linear Regression)
The data (X1, X2, X3, X4) are for each student.
X1 = score on exam #1
X2 = score on exam #2
X3 = score on exam #3
X4 = score on final exam

X1	X2	X3	X4
73	80	75	152
93	88	93	185
89	91	90	180
96	98	100	196
73	66	70	142
53	46	55	101
69	74	77	149
47	56	60	115
87	79	90	175
79	70	88	164
69	70	73	141
70	65	74	141
93	95	91	184
79	80	73	152
70	73	78	148
93	89	96	192
78	75	68	147
81	90	93	183
88	92	86	177
78	83	77	159
82	86	90	177
86	82	89	175
78	83	85	175
76	83	71	149
96	93	95	192

File names

Excel: Mlr03.xls
Minitab: Mlr03.mtp
TI-83plus/ASCII: X1 data is stored in Mlr03L1.txt
X2 data is stored in Mlr03L2.txt
X3 data is stored in Mlr03L3.txt
X4 data is stored in Mlr03L4.txt

04. **Section 10.5, Problem #5 Hollywood Movies (Multiple Linear Regression)**
The data (X1, X2, X3, X4) are for each movie
X1 = first year box office receipts/millions
X2 = total production costs/millions
X3 = total promotional costs/millions
X4 = total book sales/millions

X1	X2	X3	X4
85.10	8.50	5.10	4.70
106.30	12.90	5.80	8.80
50.20	5.20	2.10	15.10
130.60	10.70	8.40	12.20
54.80	3.10	2.90	10.60
30.30	3.50	1.20	3.50
79.40	9.20	3.70	9.70
91.00	9.00	7.60	5.90
135.40	15.10	7.70	20.80
89.30	10.20	4.50	7.90

File names	Excel: Mlr04.xls
	Minitab: Mlr04.mtp
	TI-83plus/ASCII:X1 data is stored in Mlr04L1.txt
	X2 data is stored in Mlr04L2.txt
	X3 data is stored in Mlr04L3.txt
	X4 data is stored in Mlr04L4.txt

05.   **Section 10.5, Problem #6 All Greens Franchise (Multiple Linear Regression)**
     The data (X1, X2, X3, X4, X5, X6) are for each franchise store.
     X1 = annual net sales/$1000
     X2 = number sq. ft./1000
     X3 = inventory/$1000
     X4 = amount spent on advertizing/$1000
     X5 = size of sales district/1000 families
     X6 = number of competing stores in district

X1	X2	X3	X4	X5	X6
231.00	3.00	294.00	8.20	8.20	11.00
156.00	2.20	232.00	6.90	4.10	12.00
10.00	0.50	149.00	3.00	4.30	15.00
519.00	5.50	600.00	12.00	16.10	1.00
437.00	4.40	567.00	10.60	14.10	5.00
487.00	4.80	571.00	11.80	12.70	4.00
299.00	3.10	512.00	8.10	10.10	10.00
195.00	2.50	347.00	7.70	8.40	12.00
20.00	1.20	212.00	3.30	2.10	15.00
68.00	0.60	102.00	4.90	4.70	8.00
570.00	5.40	788.00	17.40	12.30	1.00
428.00	4.20	577.00	10.50	14.00	7.00
464.00	4.70	535.00	11.30	15.00	3.00
15.00	0.60	163.00	2.50	2.50	14.00
65.00	1.20	168.00	4.70	3.30	11.00
98.00	1.60	151.00	4.60	2.70	10.00
398.00	4.30	342.00	5.50	16.00	4.00
161.00	2.60	196.00	7.20	6.30	13.00
397.00	3.80	453.00	10.40	13.90	7.00
497.00	5.30	518.00	11.50	16.30	1.00
528.00	5.60	615.00	12.30	16.00	0.00
99.00	0.80	278.00	2.80	6.50	14.00
0.50	1.10	142.00	3.10	1.60	12.00
347.00	3.60	461.00	9.60	11.30	6.00
341.00	3.50	382.00	9.80	11.50	5.00
507.00	5.10	590.00	12.00	15.70	0.00
400.00	8.60	517.00	7.00	12.00	8.00

File names   Excel: Mlr05.xls
                              Minitab: Mlr05.mtp
                              TI-83plus/ASCII: X1 data is stored in Mlr05L1.txt
                                             X2 data is stored in Mlr05L2.txt
                                             X3 data is stored in Mlr05L3.txt
                                             X4 data is stored in Mlr05L4.txt
                                             X5 data is stored in Mlr05L5.txt
                                             X6 data is stored in Mlr05L6.txt

06. Crime (Multiple Linear Regression)

This is a case study of education, crime, and police funding for small cities in ten eastern and south eastern states. The states are New Hampshire, Connecticut, Rhode Island, Maine, New York, Virginia, North Carolina, South Carolina, Georgia, and Florida.

The data (X1, X2, X3, X4, X5, X6, X7) are for each city.
X1 = total overall reported crime rate per 1million residents
X2 = reported violent crime rate per 100,000 residents
X3 = annual police funding in dollars per resident
X4 = percent of people 25 years and older that have had 4 years of high school
X5 = percent of 16 to 19 year-olds not in highschool and not   highschool graduates.
X6 = percent of 18 to 24 year-olds enrolled in college
X7 = percent of people 25 years and older with at least 4 years of college

Reference: *Life In America's Small Cities*, By G.S. Thomas

X1	X2	X3	X4	X5	X6	X7
478	184	40	74	11	31	20
494	213	32	72	11	43	18
643	347	57	70	18	16	16
341	565	31	71	11	25	19
773	327	67	72	9	29	24
603	260	25	68	8	32	15
484	325	34	68	12	24	14
546	102	33	62	13	28	11
424	38	36	69	7	25	12
548	226	31	66	9	58	15
506	137	35	60	13	21	9
819	369	30	81	4	77	36
541	109	44	66	9	37	12
491	809	32	67	11	37	16
514	29	30	65	12	35	11
371	245	16	64	10	42	14
457	118	29	64	12	21	10
437	148	36	62	7	81	27
570	387	30	59	15	31	16
432	98	23	56	15	50	15
619	608	33	46	22	24	8
357	218	35	54	14	27	13
623	254	38	54	20	22	11
547	697	44	45	26	18	8
792	827	28	57	12	23	11
799	693	35	57	9	60	18
439	448	31	61	19	14	12
867	942	39	52	17	31	10
912	1017	27	44	21	24	9
462	216	36	43	18	23	8
859	673	38	48	19	22	10
805	989	46	57	14	25	12

Data continued

X1	X2	X3	X4	X5	X6	X7
652	630	29	47	19	25	9
776	404	32	50	19	21	9
919	692	39	48	16	32	11
732	1517	44	49	13	31	14
657	879	33	72	13	13	22
1419	631	43	59	14	21	13
989	1375	22	49	9	46	13
821	1139	30	54	13	27	12
1740	3545	86	62	22	18	15
815	706	30	47	17	39	11
760	451	32	45	34	15	10
936	433	43	48	26	23	12
863	601	20	69	23	7	12
783	1024	55	42	23	23	11
715	457	44	49	18	30	12
1504	1441	37	57	15	35	13
1324	1022	82	72	22	15	16
940	1244	66	67	26	18	16

File names   Excel: Mlr06.xls
Minitab: Mlr06.mtp
TI-83plus/ASCII: X1 data is stored in Mlr06L1.txt
X2 data is stored in Mlr06L2.txt
X3 data is stored in Mlr06L3.txt
X4 data is stored in Mlr06L4.txt
X5 data is stored in Mlr06L5.txt
X6 data is stored in Mlr06L6.txt
X7 data is stored in Mlr06L7.txt

07. Health (Multiple Linear Regression)

This is a case study of public health, income, and population density for small cities in eight Midwestern states: Ohio, Indiana, Illinois, Iowa, Missouri, Nebraska, Kansas, and Oklahoma.

The data (X1, X2, X3, X4, X5) are by city.
X1 = death rate per 1000 residents
X2 = doctor availability per 100,000 residents
X3 = hospital availability per 100,000 residents
X4 = annual per capita income in thousands of dollars
X5 = population density people per square mile

Reference: *Life In America's Small Cities*, by G.S. Thomas

X1	X2	X3	X4	X5
8.0	78	284	9.1	109
9.3	68	433	8.7	144
7.5	70	739	7.2	113
8.9	96	1792	8.9	97
10.2	74	477	8.3	206
8.3	111	362	10.9	124
8.8	77	671	10.0	152
8.8	168	636	9.1	162
10.7	82	329	8.7	150
11.7	89	634	7.6	134
8.5	149	631	10.8	292
8.3	60	257	9.5	108
8.2	96	284	8.8	111
7.9	83	603	9.5	182
10.3	130	686	8.7	129
7.4	145	345	11.2	158
9.6	112	1357	9.7	186
9.3	131	544	9.6	177
10.6	80	205	9.1	127
9.7	130	1264	9.2	179
11.6	140	688	8.3	80
8.1	154	354	8.4	103
9.8	118	1632	9.4	101
7.4	94	348	9.8	117
9.4	119	370	10.4	88
11.2	153	648	9.9	78
9.1	116	366	9.2	102
10.5	97	540	10.3	95
11.9	1 76	680	8.9	80
8.4	75	345	9.6	92
5.0	134	525	10.3	126
9.8	161	870	10.4	108
9.8	111	669	9.7	77
10.8	114	452	9.6	60
10.1	142	430	10.7	71
10.9	238	822	10.3	86

Data continued

X1	X2	X3	X4	X5
9.2	78	190	10.7	93
8.3	196	867	9.6	106
7.3	125	969	10.5	162
9.4	82	499	7.7	95
9.4	125	925	10.2	91
9.8	129	353	9.9	52
3.6	84	288	8.4	110
8.4	183	718	10.4	69
10.8	119	540	9.2	57
10.1	180	668	13.0	106
9.0	82	347	8.8	40
10.0	71	345	9.2	50
11.3	118	463	7.8	35
11.3	121	728	8.2	86
12.8	68	383	7.4	57
10.0	112	316	10.4	57
6.7	109	388	8.9	94

File names   Excel: Mlr07.xls
                  Minitab: Mlr07.mtp
                  TI-83plus/ASCII: X1 data is stored in Mlr07L1.txt
                              X2 data is stored in Mlr07L2.txt
                              X3 data is stored in Mlr07L3.txt
                              X4 data is stored in Mlr07L4.txt
                              X5 data is stored in Mlr07L5.txt

08. Baseball (Multiple Linear Regression)
A random sample of major league baseball players was obtained.

The following data (X1, X2, X3, X4, X5, X6) are by player.
X1 = batting average
X2 = runs scored/times at bat
X3 = doubles/times at bat
X4 = triples/times at bat
X5 = home runs/times at bat
X6 = strike outs/times at bat
Reference: *The Baseball Encyclopedia* 9th edition, Macmillan

X1	X2	X3	X4	X5	X6
0.283	0.144	0.049	0.012	0.013	0.086
0.276	0.125	0.039	0.013	0.002	0.062
0.281	0.141	0.045	0.021	0.013	0.074
0.328	0.189	0.043	0.001	0.030	0.032
0.290	0.161	0.044	0.011	0.070	0.076
0.296	0.186	0.047	0.018	0.050	0.007
0.248	0.106	0.036	0.008	0.012	0.095
0.228	0.117	0.030	0.006	0.003	0.145
0.305	0.174	0.050	0.008	0.061	0.112
0.254	0.094	0.041	0.005	0.014	0.124
0.269	0.147	0.047	0.012	0.009	0.111
0.300	0.141	0.058	0.010	0.011	0.070
0.307	0.135	0.041	0.009	0.005	0.065
0.214	0.100	0.037	0.003	0.004	0.138
0.329	0.189	0.058	0.014	0.011	0.032
0.310	0.149	0.050	0.012	0.050	0.060
0.252	0.119	0.040	0.008	0.049	0.233
0.308	0.158	0.038	0.013	0.003	0.068
0.342	0.259	0.060	0.016	0.085	0.158
0.358	0.193	0.066	0.021	0.037	0.083
0.340	0.155	0.051	0.020	0.012	0.040
0.304	0.197	0.052	0.008	0.054	0.095
0.248	0.133	0.037	0.003	0.043	0.135
0.367	0.196	0.063	0.026	0.010	0.031
0.325	0.206	0.054	0.027	0.010	0.048
0.244	0.110	0.025	0.006	0.000	0.061
0.245	0.096	0.044	0.003	0.022	0.151
0.318	0.193	0.063	0.020	0.037	0.081
0.207	0.154	0.045	0.008	0.000	0.252
0.320	0.204	0.053	0.017	0.013	0.070
0.243	0.141	0.041	0.007	0.051	0.264
0.317	0.209	0.057	0.030	0.017	0.058
0.199	0.100	0.029	0.007	0.011	0.188
0.294	0.158	0.034	0.019	0.005	0.014
0.221	0.087	0.038	0.006	0.015	0.142
0.301	0.163	0.068	0.016	0.022	0.092
0.298	0.207	0.042	0.009	0.066	0.211
0.304	0.197	0.052	0.008	0.054	0.095

Data continued

X1	X2	X3	X4	X5	X6
0.297	0.160	0.049	0.007	0.038	0.101
0.188	0.064	0.044	0.007	0.002	0.205
0.214	0.100	0.037	0.003	0.004	0.138
0.218	0.082	0.061	0.002	0.012	0.147
0.284	0.131	0.049	0.012	0.021	0.130
0.270	0.170	0.026	0.011	0.002	0.000
0.277	0.150	0.053	0.005	0.039	0.115

File names   Excel: Mlr08.xls
                    Minitab: Mlr08.mtp
                    TI-83plus/ASCII:X1 data is stored in Mlr08L1.txt
                                    X2 data is stored in Mlr08L2.txt
                                    X3 data is stored in Mlr08L3.txt
                                    X4 data is stored in Mlr08L4.txt
                                    X5 data is stored in Mlr08L5.txt
                                    X6 data is stored in Mlr08L6.txt

09.   Basketball (Multiple Linear Regression)
A random sample of professional basketball players was obtained.

The following data (X1, X2, X3, X4, X5) are for each player.
X1 = height in feet
X2 = weight in pounds
X3 = percent of successful field goals (out of 100 attempted)
X4 = percent of successful free throws (out of 100 attempted)
X5 = average points scored per game
Reference: *The official NBA basketball Encyclopedia*, Villard Books

X1	X2	X3	X4	X5
6.8	225	0.442	0.672	9.2
6.3	180	0.435	0.797	11.7
6.4	190	0.456	0.761	15.8
6.2	180	0.416	0.651	8.6
6.9	205	0.449	0.900	23.2
6.4	225	0.431	0.780	27.4
6.3	185	0.487	0.771	9.3
6.8	235	0.469	0.750	16.0
6.9	235	0.435	0.818	4.7
6.7	210	0.480	0.825	12.5
6.9	245	0.516	0.632	20.1
6.9	245	0.493	0.757	9.1
6.3	185	0.374	0.709	8.1
6.1	185	0.424	0.782	8.6
6.2	180	0.441	0.775	20.3
6.8	220	0.503	0.880	25.0
6.5	194	0.503	0.833	19.2
7.6	225	0.425	0.571	3.3
6.3	210	0.371	0.816	11.2
7.1	240	0.504	0.714	10.5
6.8	225	0.400	0.765	10.1
7.3	263	0.482	0.655	7.2
6.4	210	0.475	0.244	13.6
6.8	235	0.428	0.728	9.0
7.2	230	0.559	0.721	24.6
6.4	190	0.441	0.757	12.6
6.6	220	0.492	0.747	5.6
6.8	210	0.402	0.739	8.7
6.1	180	0.415	0.713	7.7
6.5	235	0.492	0.742	24.1
6.4	185	0.484	0.861	11.7
6.0	175	0.387	0.721	7.7
6.0	192	0.436	0.785	9.6
7.3	263	0.482	0.655	7.2
6.1	180	0.340	0.821	12.3
6.7	240	0.516	0.728	8.9
6.4	210	0.475	0.846	13.6
5.8	160	0.412	0.813	11.2
6.9	230	0.411	0.595	2.8

Data continued

X1	X2	X3	X4	X5
7.0	245	0.407	0.573	3.2
7.3	228	0.445	0.726	9.4
5.9	155	0.291	0.707	11.9
6.2	200	0.449	0.804	15.4
6.8	235	0.546	0.784	7.4
7.0	235	0.480	0.744	18.9
5.9	105	0.359	0.839	7.9
6.1	180	0.528	0.790	12.2
5.7	185	0.352	0.701	11.0
7.1	245	0.414	0.778	2.8
5.8	180	0.425	0.872	11.8
7.4	240	0.599	0.713	17.1
6.8	225	0.482	0.701	11.6
6.8	215	0.457	0.734	5.8
7.0	230	0.435	0.764	8.3

File names   Excel: Mlr09.xls
Minitab: Mlr09.mtp
TI-83plus/ASCII: X1 data is stored in Mlr09L1.txt
                   X2 data is stored in Mlr09L2.txt
                   X3 data is stored in Mlr09L3.txt
                   X4 data is stored in Mlr09L4.txt
                   X5 data is stored in Mlr09L5.txt

10. Denver Neighborhoods (Multiple Linear Regression)
A random sample of Denver neighborhoods was obtained.

The data (X1, X2, X3, X4, X5, X6, X7) are for each neighborhood
X1 = total population (in thousands)
X2 = percentage change in population over past several years
X3 = percentage of children (under 18) in population
X4 = percentage free school lunch participation
X5 = percentage change in household income over past several years
X6 = crime rate (per 1000 population)
X7 = percentage change in crime rate over past several years
Reference: The Piton Foundation, Denver, Colorado

X1	X2	X3	X4	X5	X6	X7
6.9	1.8	30.2	58.3	27.3	84.9	-14.2
8.4	28.5	38.8	87.5	39.8	172.6	-34.1
5.7	7.8	31.7	83.5	26.0	154.2	-15.8
7.4	2.3	24.2	14.2	29.4	35.2	-13.9
8.5	-0.7	28.1	46.7	26.6	69.2	-13.9
13.8	7.2	10.4	57.9	26.2	111.0	-22.6
1.7	32.2	7.5	73.8	50.5	704.1	-40.9
3.6	7.4	30.0	61.3	26.4	69.9	4.0
8.2	10.2	12.1	41.0	11.7	65.4	-32.5
5.0	10.5	13.6	17.4	14.7	132.1	-8.1
2.1	0.3	18.3	34.4	24.2	179.9	12.3
4.2	8.1	21.3	64.9	21.7	139.9	-35.0
3.9	2.0	33.1	82.0	26.3	108.7	-2.0
4.1	10.8	38.3	83.3	32.6	123.2	-2.2
4.2	1.9	36.9	61.8	21.6	104.7	-14.2
9.4	-1.5	22.4	22.2	33.5	61.5	-32.7
3.6	-0.3	19.6	8.6	27.0	68.2	-13.4
7.6	5.5	29.1	62.8	32.2	96.9	-8.7
8.5	4.8	32.8	86.2	16.0	258.0	0.5
7.5	2.3	26.5	18.7	23.7	32.0	-0.6
4.1	17.3	41.5	78.6	23.5	127.0	-12.5
4.6	68.6	39.0	14.6	38.2	27.1	45.4
7.2	3.0	20.2	41.4	27.6	70.7	-38.2
13.4	7.1	20.4	13.9	22.5	38.3	-33.6
10.3	1.4	29.8	43.7	29.4	54.0	-10.0
9.4	4.6	36.0	78.2	29.9	101.5	-14.6
2.5	-3.3	37.6	88.5	27.5	185.9	-7.6
10.3	-0.5	31.8	57.2	27.2	61.2	-17.6
7.5	22.3	28.6	5.7	31.3	38.6	27.2
18.7	6.2	39.7	55.8	28.7	52.6	-2.9
5.1	-2.0	23.8	29.0	29.3	62.6	-10.3
3.7	19.6	12.3	77.3	32.0	207.7	-45.6
10.3	3.0	31.1	51.7	26.2	42.4	-31.9
7.3	19.2	32.9	68.1	25.2	105.2	-35.7
4.2	7.0	22.1	41.2	21.4	68.6	-8.8
2.1	5.4	27.1	60.0	23.5	157.3	6.2
2.5	2.8	20.3	29.8	24.1	58.5	-27.5
8.1	8.5	30.0	66.4	26.0	63.1	-37.4

Data continued

X1	X2	X3	X4	X5	X6	X7
10.3	-1.9	15.9	39.9	38.5	86.4	-13.5
10.5	2.8	36.4	72.3	26.0	77.5	-21.6
5.8	2.0	24.2	19.5	28.3	63.5	2.2
6.9	2.9	20.7	6.6	25.8	68.9	-2.4
9.3	4.9	34.9	82.4	18.4	102.8	-12.0
11.4	2.6	38.7	78.2	18.4	86.6	-12.8

File names   Excel: Mlr10.xls
                    Minitab: Mlr10.mtp
                    TI-83plus/ASCII: X1 data is stored in Mlr10L1.txt
                                     X2 data is stored in Mlr10L2.txt
                                     X3 data is stored in Mlr10L3.txt
                                     X4 data is stored in Mlr10L4.txt
                                     X5 data is stored in Mlr10L5.txt
                                     X6 data is stored in Mlr10L6.txt
                                     X7 data is stored in Mlr10L7.txt

11. **Chapter 10 Using Technology: U.S. Economy Case Study (Multiple Linear Regression)**
U.S. economic data 1976 to 1987
X1 = dollars/barrel crude oil
X2 = % interest on ten yr. U.S. treasury notes
X3 = foreign investments/billions of dollars
X4 = Dow Jones industrial average
X5 = gross national product/billions of dollars
X6 = purchasing power u.s. dollar (1983 base)
X7 = consumer debt/billions of dollars
Reference: *Statistical Abstract of the United States* 103rd and 109th edition

X1	X2	X3	X4	X5	X6	X7
10.90	7.61	31.00	974.90	1718.00	1.76	234.40
12.00	7.42	35.00	894.60	1918.00	1.65	263.80
12.50	8.41	42.00	820.20	2164.00	1.53	308.30
17.70	9.44	54.00	844.40	2418.00	1.38	347.50
28.10	11.46	83.00	891.40	2732.00	1.22	349.40
35.60	13.91	109.00	932.90	3053.00	1.10	366.60
31.80	13.00	125.00	884.40	3166.00	1.03	381.10
29.00	11.11	137.00	1190.30	3406.00	1.00	430.40
28.60	12.44	165.00	1178.50	3772.00	0.96	511.80
26.80	10.62	185.00	1328.20	4015.00	0.93	592.40
14.60	7.68	209.00	1792.80	4240.00	0.91	646.10
17.90	8.38	244.00	2276.00	4527.00	0.88	685.50

File names   Excel: Mlr11.xls
                    Minitab: Mlr11.mtp
                    TI-83plus/ASCII: X1 data is stored in Mlr11L1.txt
                                     X2 data is stored in Mlr11L2.txt
                                     X3 data is stored in Mlr113.txt
                                     X4 data is stored in Mlr114.txt
                                     X5 data is stored in Mlr115.txt
                                     X6 data is stored in Mlr116.txt
                                     X7 data is stored in Mlr117.txt

One-Way ANOVA
File name prefix: Owan followed by the number of the data file

01. Excavation Depth and Archaeology (One-Way ANOVA)

Four different excavation sites at an archeological area in New Mexico gave the following depths (cm) for significant archaeological discoveries.

X1 = depths at Site I
X2 = depths at Site II
X3 = depths at Site III
X4 = depths at Site IV

Reference: *Mimbres Mogollon Archaeology* by Woosley and McIntyre, Univ. of New Mexico Press

X1	X2	X3	X4
93	85	100	96
120	45	75	58
65	80	65	95
105	28	40	90
115	75	73	65
82	70	65	80
99	65	50	85
87	55	30	95
100	50	45	82
90	40	50	
78	45		
95	55		
93			
88			
110			

File names   Excel: Owan01.xls
Minitab: Owan01.mtp
TI-83plus/ASCII: X1 data is stored in Owan01L1.txt
      X2 data is stored in Owan01L2.txt
      X3 data is stored in Owan01L3.txt
      X4 data is stored in Owan01L4.txt

02. Apple Orchard Experiment (One-Way ANOVA)

Five types of root-stock were used in an apple orchard grafting experiment. The following data represent the extension growth (cm) after four years.

X1 = extension growth for type I
X2 = extension growth for type II
X3 = extension growth for type III
X4 = extension growth for type IV
X5 = extension growth for type V
Reference: S.C. Pearce, University of Kent at Canterbury, England

X1	X2	X3	X4	X5
2569	2074	2505	2838	1532
2928	2885	2315	2351	2552
2865	3378	2667	3001	3083
3844	3906	2390	2439	2330
3027	2782	3021	2199	2079
2336	3018	3085	3318	3366
3211	3383	3308	3601	2416
3037	3447	3231	3291	3100

File names	Excel: Owan02.xls
	Minitab: Owan02.mtp
	TI-83plus/ASCII:X1 data is stored in Owan02L1.txt
	X2 data is stored in Owan02L2.txt
	X3 data is stored in Owan02L3.txt
	X4 data is stored in Owan02L4.txt
	X5 data is stored in Owan02L5.txt

03. Red Dye Number 40 (One-Way ANOVA)

S.W. Laagakos and F. Mosteller of Harvard University fed mice different doses of red dye number 40 and recorded the time of death in weeks. Results for female mice, dosage and time of death are shown in the data

$X1$ = time of death for control group

$X2$ = time of death for group with low dosage

$X3$ = time of death for group with medium dosage

$X4$ = time of death for group with high dosage

Reference: *Journal Natl. Cancer Inst.*, Vol. 66, p 197-212

X1	X2	X3	X4
70	49	30	34
77	60	37	36
83	63	56	48
87	67	65	48
92	70	76	65
93	74	83	91
100	77	87	98
102	80	90	102
102	89	94	
103		97	
96			

File names	Excel: Owan03.xls
	Minitab: Owan03.mtp
	TI-83plus/ASCII: X1 data is stored in Owan03L1.txt
	X2 data is stored in Owan03L2.txt
	X3 data is stored in Owan03L3.txt
	X4 data is stored in Owan03L4.txt

04. **Business Startup Costs (One-Way ANOVA)**

The following data represent business startup costs (thousands of dollars) for shops.

X1 = startup costs for pizza
X2 = startup costs for baker/donuts
X3 = startup costs for shoe stores
X4 = startup costs for gift shops
X5 = startup costs for pet stores
Reference: *Business Opportunities Handbook*

X1	X2	X3	X4	X5
80	150	48	100	25
125	40	35	96	80
35	120	95	35	30
58	75	45	99	35
110	160	75	75	30
140	60	115	150	28
97	45	42	45	20
50	100	78	100	75
65	86	65	120	48
79	87	125	50	20
35	90			50
85				75
120				55
				60
				85
				110

File names	
	Excel: Owan04.xls
	Minitab: Owan04.mtp
	TI-83plus/ASCII: X1 data is stored in Owan04L1.txt
	X2 data is stored in Owan04L2.txt
	X3 data is stored in Owan04L3.txt
	X4 data is stored in Owan04L4.txt
	X5 data is stored in Owan04L5.txt

05. Weights of Football Players (One-Way ANOVA)

The following data represent weights (pounds) of a random sample of professional football players on the following teams.

X1 = weights of players for the Dallas Cowboys
X2 = weights of players for the Green Bay Packers
X3 = weights of players for the Denver Broncos
X4 = weights of players for the Miami Dolphins
X5 = weights of players for the San Francisco Forty Niners

Reference: *The Sports Encyclopedia Pro Football*

X1	X2	X3	X4	X5
250	260	270	260	247
255	271	250	255	249
255	258	281	265	255
264	263	273	257	247
250	267	257	268	244
265	254	264	263	245
245	255	233	247	249
252	250	254	253	260
266	248	268	251	217
246	240	252	252	208
251	254	256	266	228
263	275	265	264	253
248	270	252	210	249
228	225	256	236	223
221	222	235	225	221
223	230	216	230	228
220	225	241	232	271

File names	Excel: Owan05.xls
	Minitab: Owan05.mtp
	TI-83plus/ASCII: X1 data is stored in Owan05L1.txt
	X2 data is stored in Owan05L2.txt
	X3 data is stored in Owan05L3.txt
	X4 data is stored in Owan05L4.txt
	X5 data is stored in Owan05L5.txt

Two-Way ANOVA
File name prefix: Twan followed by the number of the data file

01. Political Affiliation (Two-Way ANOVA)
Response: Percent of voters for a recent National Election
Factor 1: counties in Montana
Factor 2: political affiliation
Reference: *County and City Data Book* U.S. Dept. of Commerce

County	Democrat	Republican
Jefferson	33.5	36.5
Lewis/Clark	42.5	35.7
Powder River	22.3	47.3
Stillwataer	32.4	38.2
Sweet Grass	21.9	48.8
Yellowstone	35.7	40.4

File names   Excel: Twan01.xls
Minitab: Twan01.mtp
ASCII: Twan01.txt

02. Density of Artifacts (Two-Way ANOVA)
Response: Average density of artifacts, number of artifacts per cubic meter
Factor 1: archeological excavation site
Factor 2: depth (cm) at which artifacts are found
Reference: Museum of New Mexico, Laboratory of Antrhopology

Site	50-100	101-150	151-200
I	3.8	4.9	3.4
II	4.1	4.1	2.7
III	2.9	3.8	4.4
IV	3.5	3.3	3
V	5.2	5.1	5.3
VI	3.6	4.6	4.5
VII	4.5	3.7	2.8

File names   Excel: Twan02.xls
Minitab: Twan02.mtp
ASCII: Twan02.txt

03. Spruce Moth Traps (Two-Way ANOVA)

Response: number of spruce moths found in trap after 48 hours
Factor 1: Location of trap in tree (top branches, middle branches, lower branches, ground)
Factor 2: Type of lure in trap (scent, sugar, chemical)

Location	Scent	Sugar	Chemical
Top	28	35	32
	19	22	29
	32	33	16
	15	21	18
	13	17	20
Middle	39	36	37
	12	38	40
	42	44	18
	25	27	28
	21	22	36
Lower	44	42	35
	21	17	39
	38	31	41
	32	29	31
	29	37	34
Ground	17	18	22
	12	27	25
	23	15	14
	19	29	16
	14	16	19

File names

Excel: Twan03.xls
Minitab: Twan03.mtp
ASCII: Twan03.txt

04. Advertising in Local Newspapers (Two-Way ANOVA)

Response: Number of inquiries resulting from advertisement
Factor 1: day of week (Monday through Friday)
Factor 2: section of newspaper (news, business, sports)

Day	News	Business	Sports
Monday	11	10	4
	8	12	3
	6	13	5
	8	11	6
Tuesday	9	7	5
	10	8	8
	10	11	6
	12	9	7
Wednesday	8	7	5
	9	8	9
	9	10	7
	11	9	6
Thrusday	4	9	7
	5	6	6
	3	8	6
	5	8	5
Friday	13	10	12
	12	9	10
	11	9	11
	14	8	12

File names   Excel: Twan04.xls
Minitab: Twan04.mtp
ASCII: Twan04.txt

05. Prehistoric Ceramic Sherds (Two-Way ANOVA)

Response: number of sherds

Factor 1: region of archaeological excavation

Factor 2: type of ceramic sherd (three circle red on white, Mogollon red on brown, Mimbres corrugated, bold face black on white)

Reference: *Mimbres Mogollon Archaeology* by Woosley and McIntyre, University of New Mexico Press

Region	Red on White	Mogollon	Mimbres	Bold Face
I	68	49	78	95
	33	61	53	122
	45	52	35	133
II	59	71	54	78
	43	41	51	98
	37	63	69	89
III	54	67	44	41
	91	46	76	29
	81	51	55	63
IV	55	45	78	56
	53	58	49	81
	42	72	46	35
V	27	47	41	46
	31	39	36	22
	38	53	25	26

File names

Excel: Twan05.xls
Minitab: Twan05.mtp
ASCII: Twan05.txt